Graded exercises in
Advanced level mathematics

Graded exercises in statistics

Richard Norris

CAMBRIDGE
UNIVERSITY PRESS

CAMBRIDGE UNIVERSITY PRESS
Cambridge, New York, Melbourne, Madrid, Cape Town, Singapore, São Paulo

Cambridge University Press
The Edinburgh Building, Cambridge CB2 8RU, UK

Published in the United States of America by Cambridge University Press, New York

www.cambridge.org
Information on this title: www.cambridge.org/9780521653992

© Cambridge University Press 2000

First published 2000
Re-issued in this digitally printed version 2008

A catalogue record for this publication is available from the British Library

ISBN 978-0-521-65399-2 paperback

ACKNOWLEDGEMENTS
The author and publishers are grateful to the following Examination Boards for permission
to reproduce questions from past examination papers:
AEB   Associated Examining Board
Edexel   Edexel Foundation
MEI   Oxford and Cambridge Schools Examination Board
OCSEB   Oxford and Cambridge Schools Examination Board
UCLEAC   University of London Examinations and Assessment Committee
UCLES   University of Cambridge Local Examinations Syndicate

Contents

Introduction

This is *not* a textbook. It provides sets of questions, graded from the very simple to the most difficult, to help you build up your confidence when you come to tackle examination questions. The questions have been designed to be student-friendly, the subject matter being drawn, as far as is possible, from everyday experience. Memorising formulae will save time in examinations. A good, and painless, way to learn them is to write them down every time you do a question.

Each section has four groups of questions.

Basic questions will give you routine practice. They will help you to remember formulae and methods of answering questions.

Intermediate questions offer a greater challenge. Several will require more than one stage of working and you may have to make a plan to solve the problem. The style of these questions will lead you into the more complex ideas which are needed in the examination.

Advanced questions are even more challenging, and are designed to be as close as possible to the level of difficulty of A Level papers. At the end of each section, you will find one or two questions from past papers.

Revision questions are extra, straightforward questions, so that you can go back over the ideas of each section.

Answers are given for all questions (except proofs, definitions and some diagrams). Questions marked with an asterisk (*) have full **worked solutions** and are intended to help you understand key topics.

Acknowledgements

I would like to acknowledge the enormous help given to me by Rodney Golton. His enthusiasm and expertise have enabled me to become (nearly) computer literate. My grateful thanks also must go to my wife Mary for her patience and constant encouragement.

Richard Norris

1

Position and dispersion, display and description

1.1 Frequency distributions, percentiles, standard deviation, skewness

$$\text{Standard deviation} = \sqrt{\frac{\sum (x - \bar{x})^2}{\sum f}} = \sqrt{\frac{\sum x^2 f}{\sum f} - (\bar{x})^2}$$

Coefficients of skewness:

$$\text{Pearson} = \frac{\text{mean} - \text{mode}}{\sigma} = \frac{3(\text{mean} - \text{median})}{\sigma}$$

$$\text{Quartile} = \frac{Q_1 - 2Q_2 + Q_3}{Q_3 - Q_1}$$

where Q_1 is the lower quartile, Q_2 is the median, Q_3 is the upper quartile

Basic

All median values have been calculated, unless otherwise requested, using interpolation.

1 For each set of data given below,

    (i) construct a frequency distribution,
   (ii) write down the mode and median,
  (iii) calculate, where possible, the mean.

(a) The number of TV sets in each of 30 households

    1 2 2 4 2 1 1 1 2 1 3 1 1 1 3
    1 2 2 1 2 1 0 3 3 1 2 1 1 0 1

(b) The number of children in each of 25 families

    2 2 5 1 2 3 1 2 4 2 2 1 3
    2 4 1 2 2 0 3 2 1 2 2 3

(c) The number of cars owned by each of 50 families

1 1 2 1 1 1 2 1 2 1 0 1 1 2 3 1 1 1 2 2 1 1 3 1 1
2 1 0 2 1 2 1 2 1 1 4 1 3 1 1 1 1 1 2 1 1 3 2 1 1

(d) The examination grades achieved by a class of 30 students

B C C E A C B B D D D C B C C
C A C B E A D C B E C B E C D

2 Various frequency distributions are shown below. For each one,
  (i) write down the mode,
  (ii) write down the median,
  (iii) calculate the mean where this is possible.

(a) The number of pairs of shoes owned by 30 students

Number of pairs	3	4	5	6	7
Frequency	6	10	12	1	1

(b) From a row of 40 cars, parked in a street, the numbers displaying the registration letters shown

Letter	J	K	L	M	N	P	R
Frequency	12	10	7	4	3	3	1

(c) The number of eggs ordered for breakfast by 50 of a café's customers

Number of eggs	0	1	2	3	4
Frequency	6	17	13	10	4

*3 A fair die is thrown 100 times, giving the following record:

Score	1	2	3	4	5	6
Frequency	17	15	16	18	16	18

Find the mean and median scores from this distribution.

4 For each of the frequency distributions given, find:

(i) the mode and median, (ii) the mean, (iii) the standard deviation.

(a)
x	2	3	4	5	6	7
f	4	9	16	14	11	6

(b)
x	0	2	4	6	8	10
f	7	16	28	32	15	2

(c)
x	5	10	15	20	25	30	35	40
f	3	7	12	20	28	31	28	21

(d)
x	2.3	3.0	5.1	5.8	6.7	7.4	9.2
f	5	8	14	22	13	5	3

5 A rugby team plays 42 matches in the course of a season. The points it scores in each game are listed below:

18	12	25	14	10	28	33	17	20	6	21	15	8	11
30	24	16	12	3	27	18	7	13	18	21	26	14	9
10	19	23	27	15	10	5	38	30	24	16	11	26	20

(a) Write down the modal and median groups, using classes $5 \leq s < 10$, $10 \leq s < 15, \ldots, 35 \leq s < 40$.

(b) Calculate the average score per match and the standard deviation of the scores.

6 For each of the frequency distributions given below, calculate the standard deviation

(i) using $\sigma = \sqrt{\left\{ \dfrac{\sum f_i x_i^2}{n} - \bar{x}^2 \right\}}$, (ii) using $\sqrt{\left\{ \dfrac{\sum f_i (x_i - \bar{x})^2}{n} \right\}}$.

(a)
x	1	2	3	4	5	6
f	10	14	20	15	11	5

(b)
x	0	2	4	6	8	10
f	1	3	8	12	16	20

(c)
x	-2	-1	0	1	2	3	4
f	3	7	9	14	20	16	11

7 From each set of continuous raw data below:

(i) draw up a grouped frequency distribution, using the class boundaries shown,
(ii) find the modal and median groups,
(iii) calculate an estimate for the mean, using the mid-range values.

(a) The lengths in centimetres, to the nearest centimetre, of 50 carrots

15	22	21	11	18	14	18	16	15	13	14	15	13	19
20	21	18	17	14	11	15	15	16	14	19	15	17	16
13	14	15	17	12	14	16	16	18	20	15	12	13	18
12	15	14	13	11	14	16	15						

Class boundaries:
$10.5 \leq l < 12.5, 12.5 \leq l < 14.5, \ldots, 20.5 \leq l < 22.5$.

(b) The masses, in grams to the nearest 0.1 gram, of 40 Christmas cards

10.4	6.3	8.7	7.8	8.8	9.2	6.9	11.1	14.0	12.2
11.4	9.4	8.6	7.1	8.2	11.0	9.3	8.8	10.1	10.2
7.7	8.6	9.7	10.9	13.6	9.8	8.9	9.2	10.8	9.4
6.2	8.8	9.4	9.9	11.1	10.1	11.8	11.2	10.1	8.3

Class boundaries:
$5.5 \leq c < 7.5, 7.5 \leq c < 8.5, \ldots, 10.5 \leq c < 11.5, 11.5 \leq c < 15.5$.

(c) The masses of 20 children in a classroom, in kilograms to the nearest 100 g

45.7 63.2 51.1 58.7 60.3 49.8 52.4 56.6 53.7 54.0
56.9 59.3 57.2 54.2 53.8 51.7 55.8 53.3 52.8 55.4

Class boundaries:
$45.5 \leq m < 48.5, \ldots, 57.5 \leq m < 60.5, 60.5 \leq m < 70.5$.

8 The grouped frequency distribution below represents the results of 100 students' efforts in a mathematics test where the maximum mark was 100. Find the modal and median groups and calculate estimates for the mean and standard deviation.

Mark group	0–19	20–39	40–49	50–59	60–69	70–79	80–100
Frequency	7	13	17	28	15	12	8

*9 Find estimates for the mean and variance of the grouped frequency distribution given below.

x	$1 \leq x < 3$	$3 \leq x < 5$	$5 \leq x < 7$	$7 \leq x < 10$
f	9	22	27	17

10 A set of data has a mean of 5 and standard deviation 1.3. The mode and median are 6 and 5.2 respectively. Calculate the two versions of Pearson's coefficient of skewness.

11 For the two frequency distributions given, find:
(i) the mean and standard deviation,
(ii) the three quartiles,
(iii) Pearson's coefficient of skewness and the quartile coefficient of skewness.

(a)
x	1	2	3	4	5	6	7	8
f	4	9	16	14	8	7	5	4

(b)
x	0	1	2	3	4	5	6	7	8
f	1	2	4	5	7	13	22	18	10

12 Two classes of 20 pupils each take the same statistics test with results as shown.

Mark, M	$0 \leq M < 10$	$10 \leq M < 20$	$20 \leq M < 30$	$30 \leq M < 40$	$40 \leq M < 50$	$50 \leq M < 60$
Class 1	2	5	9	3	1	0
Class 2	0	2	4	7	6	1

Draw, on the same sheet of graph paper, frequency polygons, so that the results may be compared. Comment on the shapes of the polygons. Estimate the mean and standard deviation for each class.

Intermediate

1 The frequency distribution for the variable x shown below has a mean of 20 and $\sum f = 76$. Find the values of A and B.

x	5	10	15	20	25	30
f	2	6	16	A	B	10

2 The waistlines of 50 men, measured in centimetres to the nearest centimetre, are shown in the table.

97	91	79	89	107	97	84	89	76	81
79	94	102	84	76	97	114	94	81	99
127	91	86	99	81	91	79	104	97	97
76	84	89	109	112	91	97	99	86	89
94	76	81	104	99	84	91	94	94	86

From these figures:

(a) write out a frequency distribution, using the following class boundaries, for the variable W, 'the waistline length'

$$75 \leq W < 85, 85 \leq W < 95, \ldots, 105 \leq W < 115, 115 \leq W < 130$$

(b) write down the modal and median groups,
(c) find estimates for the mean and standard deviation of these waistlines.

3 The maximum heights of 100 miniature daffodil flowers are recorded below. Heights are in millimetres to the nearest millimetre.

Height, h

$120 \leq h < 125, 125 \leq h < 130, 130 \leq h < 135, 135 \leq h < 140, 140 \leq h < 145, 145 \leq h < 150$

Frequency

$A+3$	$2A+1$	$(A-4)^2$	$A+3$	A	$A-3$

A is a positive integer. You are also given that, using the mid-range values of h, 122.5, 127.5, ..., $\sum fh = 13\,285$.
Calculate (a) the value of A, (b) estimates, to three significant figures, for the mean and standard deviation of h.

4 A London taxi-driver records the fares he is paid during a single, six-day, day-shift week, as shown below.

Fare, £F	$4 \leq F < 5$	$5 \leq F < 6$	$6 \leq F < 8$	$8 \leq F < 10$	$10 \leq F < 15$
Frequency	14	48	34	14	8

Show that the average fare is about £6.70. How many extra fares in the £8–£10 group would raise the average fare to £7, the others being unchanged? Give this answer to the nearest whole number.

5 Calculate the mean and variance of the following sets of data.

(a)

x	1	3	4	6	9	12	13	16	20
Frequency	3	7	12	20	25	18	8	5	2

(b)

x	$4 < x \le 7$	$7 < x \le 10$	$10 < x \le 13$	$13 < x \le 16$
Frequency	21	49	68	22

6 A well-known dog-breeder keeps a record of the number of pups per litter born to her bitches.

Number of pups	2	3	4	5	6	7
Frequency	9	13	18	15	12	8

Calculate a non-zero value of Pearson's coefficient of skewness for these figures.

7 A car dealer sold 84 second-hand cars over a period of several months. The prices of the cars are summarised in the table.

Selling price, £P	$4000 \le P < 4500$	$4500 \le P < 5000$	$5000 \le P < 6000$	$6000 \le P < 8000$
Number sold	33	26	17	8

(a) Calculate estimates for the mean and standard deviation of these prices.
(b) Draw a cumulative-frequency curve and find from it estimates for:

  (i)   the median selling price,
  (ii) the number of cars sold for more than £5500,
  (iii) the upper quartile price,
  (iv) Pearson's coefficient of skewness.

8 The masses of the eight members of a pack of rugby forwards are shown below. They are in kilograms to the nearest 0.1 kg.

        95.7   103.4   106.1   108.4   111.6   115.2   116.6   117.5

Calculate the mean and standard deviation of these figures.

9 All the aces and kings are removed from a full pack of cards and shuffled. From these eight cards, one is selected at random, replaced, and the cards shuffled again. This is done 100 times, giving the following results:
Abbreviations: A = ace, K = king, S = spades, H = hearts,
                D = diamonds C = clubs.

Card	AS	AH	AD	AC	KS	KH	KD	KC
Frequency	11	P	12	11	12	13	14	Q

You are given that $\sum$ kings $- \sum$ aces $= 4$.

(a) Find the values of the positive integers P and Q.
(b) How many times would you expect to see a red ace?

10 A teacher asks 40 students to take a short, factual revision test in preparation for a trial A level examination. Their results are:

Mark	1	2	3	4	5	6	7	8	9	10
Frequency	0	0	0	1	1	4	8	10	11	5

For this distribution, calculate:

(a) the modal and median marks,
(b) the mean and standard deviation,
(c) Pearson's coefficient of skewness, using both the quantities found in (a) (i.e. two values).

*11 Each Monday morning, all the 60 GCSE students in a school take a short numeracy test, marked out of 5. In one particular week, the head of mathematics disguises the results as follows:

Mark	0	1	2	3	4	5
Frequency	$x-2$	x	x^2	$(x+1)^2$	$2x$	$x+1$

She gives these figures to one of her sixth form mathematics classes and asks them to work out:

(a) the value of x, a positive integer,
(b) the mean and standard deviation of the marks.

What answers should the class have produced?

12 Two cubical dice are thrown together once and the scores are multiplied. This is done 144 times. The frequencies of the scores 3, 4, 6, 12 and 25 are noted.

Score	3	4	6	12	25
Frequency	5	12	7	7	7

For two fair dice, what frequences would you have expected? From the results above, can you say anything about the fairness of the two dice used?

13 An author, struggling to meet a deadline for his book to be completed, counts the number of pages he writes per day for the final 30 days. His record is given below.

No. of pages per day	16–18	19–21	22–24	25–27	27–30
No. of days this was achieved	1	3	4	9	13

(a) Calculate estimates of the mean and variance of these figures.
(b) Draw a cumulative-frequency curve to estimate the median number of pages written per day.
(c) Calculate a value for Pearson's coefficient of skewness.

14 A school enters 75 pupils for the GCSE Intermediate grade mathematics examination. Their trial examination marks are detailed below.

Mark, M	$0 \leq M < 30$	$30 \leq M < 45$	$45 \leq M < 60$	$60 \leq M < 75$	$75 \leq M < 100$
Frequency	6	13	22	27	7

(a) Calculate estimates for the mean and standard deviation of these marks.
(b) Write down the modal and median groups.
(c) Estimate the inter-quartile range from a cumulative-frequency graph.

15 A short-haul passenger jet aircraft has 125 passenger seats. The number of seats occupied during 100 flights is detailed in the table.

No. of seats	100–109	110–115	116–120	121–125
Frequency	16	24	42	18

(a) Obtain an estimate of the average number of seats, to the nearest whole number, occupied over these 100 flights.
(b) the average fare for these flights is £106. If the expenditure to the airline for the 100 flights is divided as follows:

salaries £239 000,   fuel £366 000,   food £84 000,
airport dues £250 000,   maintenance £110 000,

calculate an estimate for the percentage profit made by the airline over these 100 flights.

Advanced

1 The frequency distribution

x	A	$2A$	$3A$	$4A$	$5A$	$6A$
f	2	1	1	1	1	1

where A is a positive whole number, has a variance of 160. Find the value of A and the mean and median of the distribution.

2 Four A level classes of 15 pupils each sit a trial statistics examination. Their marks, out of a total of 99, are shown below.

```
46  31  74  68  42  54  14  61  83  48  37  26   8  64  57
93  72  53  59  38  16  88  75  56  46  66  45  61  54  27
27  44  63  58  43  81  64  67  36  49  50  76  38  47  55
77  62  53  40  71  60  58  45  42  34  46  40  59  42  29
```

(a) Draw up a grouped frequency distribution using classes 0–9, 10–19, ..., 90–99.
(b) Calculate the mean and standard deviation for these marks.
(c) The marks are to be scaled using the relation $Y = 5 + 2X$, where X is the original mark. Find the mean and standard deviation of the scaled marks.
(d) Write down the maximum and minimum values of Y.
(e) If the scaling notation is written as $Y = pX + q$, and the mean and standard deviation of Y are chosen as 100 and 20 respectively, what, to three significant figures, are the values of p and q?

3 A volunteer fund-raiser for a hockey club makes telephone calls to friends and local companies to try to raise money for an overseas tour. The lengths of the calls, in seconds, to the nearest second, are recorded, and a frequency distribution for 100 calls, made on a mobile telephone, is shown.

Time, t	$95 \leq t < 125$	$125 \leq t < 155$	$155 \leq t < 185$	$185 \leq t < 215$	$215 \leq t < 245$
Frequency	14	22	28	21	15

The cost of the calls, by arrangement with the local telephone office, is calculated according to the following scale:

first minute 2.5p per 10 seconds or part of a second,
second minute 2.0p per 10 seconds or part,
thereafter 1.0p per 10 seconds or part.

Use the mid-range values of the time intervals to calculate an estimate for the total cost of these calls. Calculate an estimate for the average length of a call and find the median length from a cumulative frequency graph.

4 The cost of a telephone bill, for local calls, issued by the Sweetalk telephone company, is made up as follows:

(i) for each three-month period there is a standing charge of £18,
(ii) the length of every call is rounded up to the nearest minute,
(iii) a standard-rate call costs 5p per minute and a cheap-rate call 3p per minute.

In a three-month period, a garrulous statistician makes 100 calls on this tariff, as shown in the table below.

Time (minutes)	5	10	12	14	20
No. of calls: standard rate	8	7	15	7	6
No. of calls: cheap rate	5	13	21	8	10

Calculate:

(a) the total cost of these calls,
(b) the average length, to the nearest second, of a call,
(c) the median length of a standard-rate call,
(d) the standard deviation for the times of the cheap-rate calls in this distribution.

5   A trainer keeps a record of the times taken by one of his horses to run a measured mile on the stable gallops. The horse is timed for 30 runs, with the following results, given in seconds to the nearest second.

88   91   87   85   84   90   88   83   87   86   91   85   88   89   82
90   87   85   83   84   86   85   81   86   85   87   82   86   84   87

Using class intervals $80 \le t < 82$, $82 \le t < 84, \ldots, 90 \le t < 92$, draw up a frequency distribution for these times and so find estimates for the mean and standard deviation. A second horse from the same stable is timed on the same days for the same 30 runs. The mean and standard deviation for these times are 83.2 and 6.13 respectively. The trainer is due to enter one of these horses in a mile race in three weeks' time. Outline briefly any arguments he might consider for each horse.

*6   A diamond prospector dreams she is on a flat plain and that she finds large numbers of gem-quality diamonds loose on the surface! In her dream, she marks off 100 equal areas, and counts the number of diamonds she finds in each, with the following results.

No. of diamonds	0	1	2	3	4	5	6
No. of areas		2	10	13	19	27	21	8

Calculate:

(a) the quartile coefficient of skewness,
(b) Pearson's coefficient of skewness.

(Use a cumulative-frequency curve to find the three quartiles for this distribution.)

7   A fair die and a biased die are each rolled 100 times. The probabilities of the biased die showing the scores 1 to 6 are given as:

Score	1	2	3	4	5	6
p(that score)	$\frac{1}{8}$	$\frac{1}{4}$	$\frac{1}{5}$	$\frac{1}{8}$	$\frac{1}{10}$	$\frac{1}{5}$

Calculate the frequencies of the scores you would expect from the biased die and so compare the means and standard deviations of the scores of the two dice.

8 The average weekly incomes, in £, of households in 11 regions of the United Kingdom are given below.

$$255.8 \quad 252.0 \quad 270.6 \quad 298.4 \quad 362.3 \quad 297.2$$
$$266.8 \quad 261.7 \quad 247.1 \quad 259.1 \quad 220.6$$

(a) Find the median and the upper and lower quartiles.
(b) On graph paper, draw a whisker plot to represent these data.
(c) Describe the skewness of this distribution and identify a possible outlier.
(d) Calculate the mean and standard deviation for these data.

Further investigations suggested that the £362.3 value could be incorrect and should in fact be £326.3.

(e) Without carrying out any further calculations indicate what effect this change would have on

(i) the standard deviation,   (ii) the inter-quartile range.

[ULEAC]

9 A random sample of size 1000 was selected from the persons listed in a residential telephone directory of a city in England. The number of letters in each surname was counted and the distribution of the length of the surnames is given in the table below.

Length of surname	3	4	5	6	7	8	9	10	11	12
Frequency		8	63	207	247	220	128	75	37	12	3

(i) Represent this distribution by a frequency polygon.
(ii) Find the median and the inter-quartile range of the length of the surnames shown in the table.
(iii) Calculate the mean length of the surnames shown in the table.
(iv) State with a reason whether the people in this sample are likely to be representative of all the people in the city.

[UCLES]

Revision

1 For each set of data below,
(i) construct a frequency distribution,
(ii) calculate the mean and standard deviation.

(a) The number of wrist watches owned by 45 young people

```
1 2 1 0 1 2 3 5 2 1 1 2 1 0 3
4 2 1 1 2 1 3 1 2 1 2 4 1 2 1
3 1 1 2 3 1 2 1 1 2 2 1 1 1 2
```

(b) The number of wickets taken by an international opening bowler in 30 cricket test match innings

```
1 2 1 0 3 5 4 2 3 1 0 0 2 1 1
0 3 2 1 1 3 1 0 1 6 2 0 2 2 1
```

2 A poker player keeps a record of the number of 'pairs' he is dealt in each hand. For 25 successive deals the record is

```
1 1 0 2 1 1 0 2 0 1 2 2 0
2 1 2 1 0 0 1 0 1 2 2 1
```

(a) Write down the frequency distribution for these results.
(b) Draw a cumulative-frequency graph and from it find the median number of pairs dealt.
(c) Calculate the average number of pairs per hand dealt to the player.

3 The masses of 50 sumo wrestlers are recorded just before an important tournament and are displayed below, in kilograms to the nearest kilogram.

Mass, M

$105 \leq M < 120$	$120 \leq M < 130$	$130 \leq M < 140$	$140 \leq M < 150$	$150 \leq M < 170$
Frequency				
4	14	21	8	3

Calculate estimates for the mean and standard deviation of the masses of these wrestlers and find the median mass from a cumulative-frequency graph.

4 Use a coding method to calculate the mean of each of the following sets of data. Give the answers to the accuracy required.

(a)

x	0.003	0.004	0.005	0.006	0.007
f	14	17	13	6	5

Three significant figures

(b)

x	4025	4019	4021	4030	3995
f	1	1	2	4	5

Nearest whole number

(c)

x	2.0008	2.0003	2.0007	2.0010	2.0005
f	2	4	1	1	3

All the figures on your calculator

5   Calculate both versions of Pearson's coefficient of skewness for the following sets of data. Use the mode and median as whole numbers from the table.

(a)
x	1	2	3	4	5	6
f	25	23	18	13	7	4

(b)
x	1	2	3	4	5	6
f	3	5	37	32	15	8

*6   For the frequency distribution below, calculate:
(a) the mean, median and mode,
(b) the variance,
(c) the quartile coefficient of skewness.

x	2	4	6	8	10	12	14
f	3	5	11	16	40	56	19

7   Over a period of 100 days in the peak summer season, a seaside hotel manager keeps a record of the number of people staying in her establishment.

No. staying	41–45	46–50	51–55	56–60	61–65
No. of days	5	14	23	30	28

Calculate:
(a) estimates of the mean and standard deviation for these figures,
(b) Pearson's coefficient of skewness (assume median = mode = 58).

8   For each of the following frequency distributions, calculate:
(i) the mean,   (ii) the variance,   (iii) the median,   (iv) the mode.

(a)
x	6	10	15	18	22	27	30
f	6	18	27	36	31	21	11

(b)
x	2.5	2.7	3.1	3.6	3.8	4.2	4.7
f	3	14	28	23	8	3	1

9   Calculate, for the frequency distribution shown, estimates for:
(a) the mean,   (b) the variance,
(c) the quartile coefficient of skewness.

x	1	2	3	4	5
f	15	39	56	64	26

10   A professional darts player, at practice, keeps a record of his scores with each dart, when he is aiming for the 'treble twenty', as follows.

Score	1	5	20	40	60
Frequency	5	7	90	42	156

(a) Calculate the mean and standard deviation of his scores.
(b) Show, using interpolation, that his median score is between 40 and 41.

11 Pumpkin seeds are sold in packets containing eight seeds. The company marketing these seeds, in order to make a claim about their germination level, plants the contents of 100 packets in adjacent rows in a seed-bed and counts the number which germinate in each row. The results are:

No. germinating	0	1	2	3	4	5	6	7	8
Frequency	0	0	1	2	2	4	19	37	35

What would be a reasonable claim for the percentage germination of the eight seeds in a packet?

12 The petrol consumption, in kilometres per litre, of 50 examples of the same model of car was tested by the manufacturers as detailed below.

Consumption (km/l)	$11.0 \leq P < 11.7$	$11.7 \leq P < 12.4$	$12.4 \leq P < 13.1$	$13.1 \leq P < 13.8$
Frequency	8	13	17	12

If the manufacturers claimed that the consumption of this model was 12.8 km/l, would this be an overstatement?

1.2   Charts, diagrams, histograms

Basic

1 A school has a budget of £2 million for a particular academic year. This sum is allocated to four areas: salaries 60%, maintenance 20%, grounds 15%, books and equipment 5%. Draw an accurate pie chart to illustrate these figures. Write in the angles for each sector. How much is spent on books and equipment?

2 A firm manufactures four products A, B, C and D. The quantities of each produced in a year are in the ratio 1 : 2 : 3 : 3 respectively.

(a) Draw an accurate pie chart to illustrate this output, showing all the sector angles.
(b) If the value of product B was £68 000 in this year, what was the value of (i) product C, (ii) the total output?

*3 A univerity college has a budget
    of £1 800 000, to be spent as
    detailed in the pie chart, which is
    not to scale. Complete the chart,
    calculating all the angles and the
    separate sums spent on the five
    listed items.

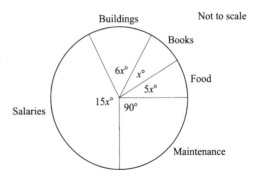

4 Draw a separate bar chart to illustrate each of the following sets of
   data.

   (a) The annual tonnage of ore from two copper mines for five
       consecutive years (the figures are in thousands of tonnes
       (1000 kg = 1 tonne) of ore)

	Year 1	Year 2	Year 3	Year 4	Year 5
Mine 1	24.2	31.4	28.7	25.1	22.3
Mine 2	21.5	24.4	23.6	20.4	17.1

   (b) The number of votes cast for two political parties in four local
       elections

Election	1	2	3	4
Party 1	4500	4200	4000	3400
Party 2	2800	2900	3300	4000

   (c) The average number of pupils in three houses in a boarding school
       over five years

	Year 1	Year 2	Year 3	Year 4	Year 5
Red house	63	65	64	69	71
White house	70	73	71	76	79
Blue house	68	68	68	72	76

5 Draw separate bar charts for the two sets of figures below.

   (a) The monthly output, in hundreds of tonnes (1 tonne = 1000 kg)
       of steel, from three blast furnaces, for three consecutive months,
       including the total output

	M_1	M_2	M_3
F_1	340	310	330
F_2	290	320	330
F_3	140	140	190

(b) The sales of three categories of lorry in two consecutive years

	Y_1	Y_2
L_1	340	345
L_2	260	290
L_3	280	255

6 Draw a stem-and-leaf diagram for each of the two sets of data given. Remember to show a key and use the required class intervals.

(a) The masses, to the nearest gram, of 30 large eggs

67 58 64 66 62 65 61 64 59 61 62 63 59 67 62
64 66 63 65 65 64 66 59 60 61 66 64 65 63 62

Class intervals: 57–59, 60–62, 63–65, 66–68.

(b) The times, in seconds, to the nearest 0.1 second, taken by 20 athletes to run a 100 m race

11.2 11.1 10.9 11.3 11.0 10.8 11.2 11.1 11.3 11.2
11.4 11.0 11.3 10.9 11.2 11.2 11.4 11.3 11.2 11.1

Class intervals:
$10.8 \leq t < 10.9$, $11.0 \leq t < 11.1$, $11.2 \leq t < 11.3$, $11.4 \leq t < 11.5$.

7 Represent each of the following sets of data by (i) a block diagram, (ii) a bar chart, (iii) a pictogram.

(a) The number of goals scored by a team per match, in a series of 30 randomly selected league hockey matches

Number of goals	0	1	2	3	4	5
Frequency	2	4	9	7	5	3

(b) The number of left-handed pupils per class in 100 randomly selected classes from all over the country, each class containing 25 pupils

Number left-handed	0	1	2	3	4
Frequency	2	13	26	38	21

8 Twenty random numbers, generated from a pocket calculator, are shown below.

0.528   0.728   0.293   0.456   0.630   0.964   0.509   0.564   0.717   0.413
0.865   0.050   0.817   0.761   0.872   0.449   0.458   0.011   0.962   0.054

Display these on a stem-and-leaf diagram, using the key 6|30 means 0.630.

9 Construct frequency polygons and calculate Pearson's coefficient of skewness for each of the sets of data given below.

(a)

x	4	5	6	7	8	9
f	2	7	14	12	6	3

(b)
x	3	6	9	12	15	18
f	6	11	10	8	7	5

(c)
x	1	2	3	4	5	6	7	8
f	15	22	19	16	12	7	6	4

(d)
x	3	4	5	6	7	8
f	12	16	22	27	29	25

10 Illustrate the data below by constructing frequency polygons. Write down the modal class and calculate Pearson's coefficient of skewness for each set of data.

(a) The lengths, L, of 100 runner beans, measured in centimetres to the nearest centimetre

L	$16 \leq L < 18$	$18 \leq L < 20$	$20 \leq L < 22$	$22 \leq L < 24$	$24 \leq L < 26$	$26 \leq L < 28$	$28 \leq L < 30$
f	4	18	24	30	12	8	4

(b) The masses, E, in grams to the nearest gram, of 200 chicken eggs

E	$53 \leq E < 55$	$55 \leq E < 57$	$57 \leq E < 59$	$59 \leq E < 61$	$61 \leq E < 63$	$63 \leq E < 65$	$65 \leq E < 67$
f	6	28	36	46	42	30	12

(c) The times, t, in seconds to the nearest 0.1 second, taken by 150 athletes to run a 110 m high hurdles race

t	$13.8 \leq t < 14.0$	$14.0 \leq t < 14.2$	$14.2 \leq t < 14.4$	$14.4 \leq t < 14.6$	$14.6 \leq t < 14.8$	$14.8 \leq t < 15.0$
f	2	4	7	69	58	10

11 Two classes of 20 pupils each take the same statistics test with results as shown.

Mark, m	$0 \leq m < 10$	$10 \leq m < 20$	$20 \leq m < 30$	$30 \leq m < 40$	$40 \leq m < 50$	$50 \leq m < 60$
Class 1	2	5	9	3	1	0
Class 2	0	2	4	7	6	1

On the same sheet of graph paper, draw frequency polygons, so that the results for each class may be compared. Comment on the shapes of the polygons. Estimate the mean and standard deviation for each class.

12 A group of 80 students sit a mental arithmetic test which is marked out of 10. The table below shows the results.

Mark	0	1	2	3	4	5	6	7	8	9	10
Frequency	0	1	3	5	9	14	18	16	9	3	2

(a) State the mode and calculate the mean mark.
(b) Draw a cumulative-frequency graph for these marks.
(c) Use the graph to find the median mark and the inter-quartile range. Compare this median with the value obtained from the table.

13 For each set of data find the three quartiles and draw a box-and-whisker plot.

(a) 13   24   7   18   4   22   9   33   28
(b) 22   14   19   7   38   16   29   47   24   36
(c) 143   120   159   151   138   146   135   144

Write down the mean, median and the upper and lower quartiles in each case.

14 Draw a box-and-whisker plot for each set of data given below. Find the inter-quartile range in each case.

(a) 1   2   3   4   5   6
(b) 2   4   6   8   10   12   14
(c) 21   25   28   33   39   46   57

15 Draw histograms to illustrate the sets of data given below and show on each diagram the corresponding frequency polygon. Write down the modal value in each case.

(a) The masses, m, of 40 men in kilograms to the nearest kilogram

m	$60 \leq m < 70$	$70 \leq m < 90$	$90 \leq m < 130$	$130 \leq m < 140$
f	5	20	10	5

(b) The speeds, v, of 500 lorries, in $km\,h^{-1}$ to the nearest whole number, passing a speed-check camera on a motorway

v	$70 \leq v < 90$	$90 \leq v < 100$	$100 \leq v < 105$	$105 \leq v < 115$	$115 \leq v < 135$	$135 \leq v < 160$
f	24	42	68	114	164	88

(c) The times t, taken by 3500 athletes to run a marathon, in minutes to the nearest minute

t	$130 \leq t < 150$	$150 \leq t < 155$	$155 \leq t < 160$	$160 \leq t < 170$	$170 \leq t < 200$
f	55	256	479	1264	1446

(d) The annual incomes, I, in thousands of pounds to the nearest £1000, of 100 families in a large town

I	$2 \leq I < 5$	$5 \leq I < 10$	$10 \leq I < 20$	$20 \leq I < 50$	$50 \leq I < 100$
f	8	26	47	16	3

Intermediate

1 The figures below show the sales of three models of a car in a year:

model 1   310   model 2   340   model 3   250

(a) Draw an accurate pie chart to show these figures. Label the sectors and angles clearly
(b) In the next year, there is an intensive sales drive which results in increases in the sales of models 1, 2 and 3 of 10%, 5% and 14%

respectively. If a new pie chart were drawn, how would it differ from the first?

2   Three garages (G_1, G_2, G_3) each sell the same four models (M_1, ..., M_4) of a car. The sales figures shown are for a period of six months.

	G_1	G_2	G_3
M_1	4	11	4
M_2	13	20	6
M_3	7	13	6
M_4	12	16	8

Draw three accurate pie charts which can be used to compare these figures. Show clearly the angles for each sector, and the radii of the three circles, if the circle for G_1 has a radius of 3 cm.

3   Two firms selling heavy printing machinery have, for a particular year, sales in millions of pounds, in four regions, as shown.

	Europe	Japan	N. America	S. America
Firm A	4.3	3.5	6.1	4.1
Firm B	7.9	7.2	11.8	9.1

Draw separate accurate pie charts for each firm so that their sales can be compared. Show the angles for each region, and state the radius of the circle for Firm B, taking the radius of the circle for Firm A as 4 cm.

4   The England cricket selectors assess four batsmen competing for one of the places in the team. The runs scored by each player in May and June are:

Batsman	1	2	3	4
May	346	296	448	377
June	294	254	312	433

Show these results on a compound bar chart. Display the total runs scored by each batsman on an accurate pie chart (angles to the nearest degree) of radius 4 cm.

5   The figures below give the percentages of pupils from three age groups, to the nearest year, present in five schools.

Age group	11–15	16–17	18+
School 1	70	24	6
School 2	42	35	23
School 3	41	39	20
School 4	54	25	21
School 5	83	12	5

Draw (a) a compound bar chart, (b) a multiple bar chart to illustrate these figures.

6 As a practice for her A level examinations, Paula takes five tests: in pure mathematics, mechanics, statistics, physics and chemistry. Her teachers forecast her results as follows:

pure mathematics 90%,   mechanics 45%,   statistics 60%, physics 30%, chemistry 75%

Show these forecasts on (a) an accurate pie chart of radius 5 cm, (b) a bar chart.

7 The thickness of the paint layer on a stationary mechanical component is checked on 25, randomly selected, finished items. The results, in millimetres to the nearest 0.01 mm, are:

0.25   0.37   0.34   0.61   0.48   0.44   0.52   0.21   0.14
0.47   0.44   0.23   0.19   0.51   0.45   0.48   0.34
0.60   0.54   0.43   0.41   0.36   0.55   0.33   0.46

Show these results on a stem-and-leaf diagram. State the class intervals and give a key to interpret the diagram.

8 The quantity of gas used (equivalent to kilowatt-hours) in a household in each of four quarters of a year is represented by a pie chart with the following angles: March to May $(3x + 3)°$, June to August $(2x - 9)°$, September to November $(2x + 6)°$, December to February $(11x)°$. Calculate these angles to the nearest degree and calculate, to the nearest whole number, the quantities of gas used, in kW h. The total consumption for the year was 18 997 kW h.

*9 Green Lane, a long straight road in the small town of Speedchester, is being used to monitor driving speeds. There is a speed limit of 60 km h^{-1} on this road. The first 40 cars gave the results shown. The figures are in km h^{-1} to the nearest 0.1 km h^{-1}.

56.3   58.4   58.7   60.4   55.7   57.1   49.5   56.6   58.2   57.4
58.5   57.8   57.3   56.6   58.6   59.4   60.8   63.3   53.9   56.0
50.6   55.4   62.7   62.1   58.4   57.6   58.8   59.3   51.1   65.3
57.9   57.5   59.8   58.2   58.3   56.8   59.4   60.5   58.0   59.2

Draw a stem-and-leaf diagram to illustrate these figures. Give a clear statement explaining your class intervals, and the key to the diagram.

10 The annual costs of running a restaurant are divided into four categories: (a) salaries, (b) cooking materials, (c) power and water, (d) rent and advertising. These are shown in the pie chart, which is not to scale. The total turnover of the restaurant was £254 000, which represented a profit of 27% on the costs. Find the outgoings for each category of the costs, and draw an accurate pie chart to illustrate these figures (radius 4 cm).

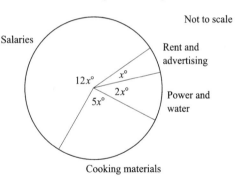

11 A rugby union scrum consists of eight members. The range of masses of the members of 40 randomly chosen scrums is as follows. The masses are in kilograms to the nearest kilogram.

$90 \leq M < 96$	$96 \leq M < 100$	$100 \leq M < 102$	$102 \leq M < 104$	$104 \leq M < 106$	$106 \leq M < 110$	$110 \leq M < 120$
2	4	9	12	7	4	2

(a) Show these results on a histogram.
(b) On the same diagram, draw a frequency polygon.
(c) On a separate diagram, draw a cumulative-frequency graph. Use this graph to estimate the median and inter-quartile range of these masses.

12 A transport company's expenditure over a year is illustrated in the pie chart. The total expenditure is £1 080 000. Find

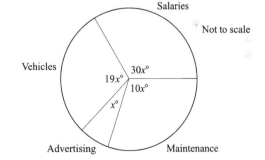

(a) the cost of each of the four items,
(b) the sizes of the four sector angles in the chart.

*13 Four local newspapers published in a city are known as *The News*, *The Phoenix*, *The Standard* and *The Monitor*. Their sales for a year, in thousands of copies, are, respectively, 5.5, 2.4, 5.8 and 4.3.

(a) Draw an accurate pie chart to show these figures. Label the sectors and show the angles clearly.
(b) In the following year, the sales figures were not available, but it is known that the sales, in the same order, were in the ratio $5:3:6:4$. Draw a second pie chart showing these sales, labelling sectors and angles as before. Comment on the changes in the sales,

when they are all expressed as percentages of the total sales, in each of the two years.

14 The ground areas of three housing estates, H_1, H_2 and H_3, are represented in the pie chart. Calculate (a) θ_2, (b) θ_3, (c) A_3.

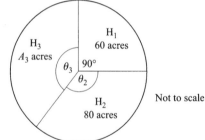

15 A company manager, faced with the possibility of having to reduce staff during a recession, compiles a table of the ages of his employees.

Age, completed years	20–30	31–35	36–40	41–45	46–50	51–60
Number in age group	5	7	18	30	16	9

Draw

(a) a cumulative frequency graph and so find estimates for the three quartiles,
(b) a histogram,
(c) a box-and-whisker plot,

to illustrate these figures.

16 For each set of data given below,

(i)   construct a histogram and superimpose the corresponding frequency polygon,
(ii) calculate estimates for the mean, median and quartiles,
(iii) draw a cumulative-frequency graph and compare estimates from it of the median and quartiles with those found in (ii).

(a) The maximum speeds of service, s, in $km\,h^{-1}$, of 40 players at an international lawn tennis tournament

s	$85 \leq s < 100$	$100 \leq s < 105$	$105 \leq s < 110$	$110 \leq s < 125$	$125 \leq s < 160$
Frequency	11	9	8	7	5

(b) The costs of 50 new home security systems sold in a city by a representative over a period of three months

Cost, £C	$750 \leq C < 850$	$850 \leq C < 950$	$950 \leq C < 1100$	$1100 \leq C < 1300$	$1300 \leq C < 1600$
Number sold	6	11	19	9	5

17 The times taken to complete a crossword at a national competition were noted for 50 competitors. Times (t) are in minutes to the nearest minute.

Time, t	$20 \leq t < 25$	$25 \leq t < 30$	$30 \leq t < 40$	$40 \leq t < 50$	$50 \leq t < 55$
No. of competitors	7	16	13	9	5

(a) Construct a histogram for these values and superimpose a frequency polygon.

(b) Calculate the angles for a pie chart (to the nearest degree) to show these results.

(c) Draw a cumulative-frequency graph to find the median time.

(d) Calculate an estimate for the mean time.

Advanced

1 (a) The quantities A, B, C, D shown in the sectors of the pie chart are to be displayed on a compound bar chart. Draw the diagram accurately, labelling each section as a percentage of the whole.

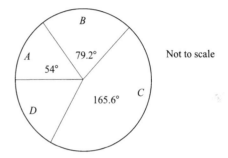

(b) Calculate the mean μ and standard deviation σ of:

(i) the four angles in the pie chart (μ_1, σ_1),

(ii) the percentages of A, B, C and D (μ_1, σ_2).

For σ_1 and σ_2 give all the figures on your calculator. What is the relation between μ_1 and μ_2?

2 The courses taken by 1440 students at a university are (i) social sciences, (ii) languages, (iii) history, (iv) mathematics and physics and (v) chemistry and biology. The numbers taking each course are in the ratio $11:5:4:3:1$ respectively.

(a) Calculate the numbers taking each course.

(b) Construct an accurate pie chart (radius 4 cm) for these figures, labelling the angles.

(c) In the following academic year, propaganda by the history and chemistry/biology departments results in the numbers taking history being doubled and the numbers taking chemistry/biology being tripled, the numbers for the other courses being unchanged. Calculate the angles of a second pie chart to show these figures, and the new ratios for the five sections.

3 A general knowledge test, marked out of 10, is given to all 77 members of the sixth form in a school. The results are displayed in the table.

Mark	0	1	2	3	4	5	6	7	8	9	10
No. scoring that mark	1	2	4	6	10	12	20	9	6	4	3

(a) Calculate the mean and standard deviation of these marks.
(b) Draw a cumulative-frequency curve and from it find the three
      quartiles.
(c) Calculate the quartile coefficient of skewness for this distribution.
      What percentage of pupils score more than 5.5 marks?

4 A firm of business suppliers has four main divisions: furniture (F),
accessories (A), paper (P) and writing instruments (W). The profits from
the divisions in one year total £900 000 and this is represented on a pie
chart. The angles of the chart are F $(x)°$, A $(x + 18)°$, P $(2x + 10)°$ and W
$(x - 18)°$.

(a) Construct an accurate pie chart of radius 4 cm for these figures and
      calculate, to the nearest pound, the profits made by each division.
(b) In the following year, the profits from each division rose by 15%,
      and a new chart was drawn. Explain why the sector angles
      remained unchanged in this new chart, but the radius had to be
      increased to compare the figures. Calculate the new radius.

5 The staff on a gold mine are divided into sections and the percentages
for each category are given below.

Manager 0.25%,   Section Officer 0.5%,   Mine Captain 3.25%,
Shift Boss 8%,   Face Worker 74%,   Administration 14%

There were 52 Mine Captains at this mine. Construct an accurate pie
chart, with radius 5 cm, and angles to the nearest degree, to illustrate
these figures. Label each sector clearly and calculate the numbers in each
group. Would a pictogram be a suitable way to display these figures?
Give a reason for your answer.

6 (a) A rectangular gardening allotment consists of 24 identical planting
      strips in which the gardener plants potatoes, lettuces, carrots, peas
      and beans. The number of rows of each vegetable planted are

Potatoes	Lettuces	Carrots	Peas	Beans
10	1	2	6	5

Construct an accurate pie chart to show these figures.
(b) In the following year, 20% more rows of potatoes were planted, but
      one row fewer of each of peas and beans. Write out a new
      frequency table and calculate the angles for a second chart.
(c) In the third year, the gardener rents two extra identical allotments,
      and plants the same vegetables in the same number of rows as in
      the first year, in each allotment. If a third pie chart were to be
      drawn, how would its angles and radius compare with those of the
      first pie chart?

7 A mathematics class of 33 pupils sits a test which is marked out of 10, giving the results shown.

Mark	0	1	2	3	4	5	6	7	8	9	10
No. scoring that mark	0	1	2	4	4	5	7	4	3	2	1

(a) Plot a cumulative-frequency graph and find from it estimates for the median and inter-quartile range.
(b) The pass mark is first set at 4 marks. To the nearest whole number, what percentage of pupils pass? If it is then felt that, taking account of previous tests, only 25% of pupils merit failure, estimate from your graph where the pass mark, to one decimal place, would have to be set.

*8 A large playing field encloses several playing pitches: two for rugby, two for hockey, four for football and one for lacrosse. The areas taken up by these sports are shown on a pie chart, the sector angles being rugby 82°, hockey 54.7° and football 164° to the nearest 0.1°. Given that the area of one of the football pitches is 0.753 hectares, calculate the area of:

(a) one hockey pitch,   (b) the lacrosse pitch,   (c) all the pitches.

Show these results on an accurate pie chart of radius 5 cm.

9 Calculate the mean and standard deviation of the numbers displayed in the stem-and-leaf diagram, according to the three different keys given below. Give answers to three significant figures.

2	3
3	1 3
4	4 4 7 9 9
5	2 4 4 6
6	2 2 5
7	8

(a) 4|7 means 0.047
(b) 4|7 means 470
(c) 4|7 means 4.07

10 The sectors of the pie chart represent the distribution of egg sizes in a batch of 1800 eggs. In grams to the nearest gram, the masses, m, of each category are as follows:

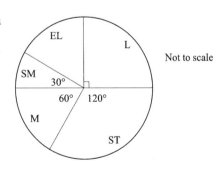

Not to scale

small (SM) $40 \leq m < 45$,
medium (M) $45 \leq m < 50$,
standard (ST) $50 \leq m < 55$,
large (L) $55 \leq m < 65$,
extra large (EL) $65 \leq m < 70$

(a) Find the number of eggs in each category.
(b) Construct a histogram to display these numbers.
(c) Find an estimate for the mean mass of an egg in the batch, to the nearest integer.

11 A mathematics teacher gives two A level classes some statistics questions for homework. He asks them to make a note of the time taken to complete the work, to the nearest minute. The times taken by the 28 pupils are as grouped below.

Time, t	$25 \leq t < 30$	$30 \leq t < 40$	$40 \leq t < 50$	$50 \leq t < 60$	$60 \leq t < 65$
Frequency	3	5	12	6	2

    (a) Draw a cumulative-frequency graph and find the median time.
    (b) Construct a histogram for these figures.
    (c) From the cumulative-frequency graph, find how many pupils took (i) more than 55 minutes, and (ii) less than 35 minutes to complete the homework.

12 The stem-and-leaf diagram shows 25 numbers, the key to the diagram being that 4|5 means 4.05.
From these figures,

    (a) construct a histogram using the six class intervals defined by the diagram,
    (b) find (i) the mean of the numbers in the diagram,

               (ii) an estimate for this mean, from the grouped frequency distribution used for the histogram.

```
1 | 2 2 3
2 | 4 5 7 7
3 | 1 1 3 4 4 5 8
4 | 2 2 5 6 8
5 | 1 3 7 8
6 | 5 9
```

13 An advertising campaign to promote electric showers consists of a mailshot which includes a prepaid postcard requesting further details. Prospective customers who return the postcard are then contacted by one of five sales staff: Gideon, Magnus, Jemma, Pandora or Muruvet. The pie charts below represent the number of potential customers contacted and the number of sales completed during a one month period.

    (a) The total number of potential customers contacted is 1000. Find, approximately, the total number of sales completed.
    (b) Describe the main features of the data revealed by the pie charts.
    (c) The manager wishes to compare the sales staff according to the number of sales completed. What type of diagram would you recommend, in place of a pie chart, so this comparison could be made easily?

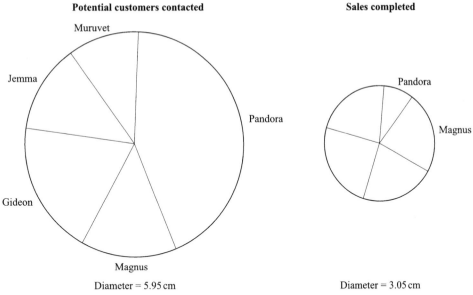

Potential customers contacted

Muruvet

Jemma

Pandora

Gideon

Magnus

Diameter = 5.95 cm

Sales completed

Pandora

Magnus

Diameter = 3.05 cm

[AEB]

14 Summarised below is the data relating to the number of minutes, to the nearest minute, that a random sample of 65 trains from Darlingborough were late arriving at a main line station.

Minutes late (0 | 2 means 2)

0	2	3	3	3	4	4	4	4	5	5	5	5	5	5	(14)
0	6	6	6	7	7	8	8	8	9						(9)
1	0	0	0	2	2	3	3	4	4	4	5				( )
1	6	6	7	7	8	8	8	9	9						( )
2	1	2	2	3	3	3	3	4							( )
2	6														( )
3	3	4	4	5											(4)
3	6	8													(2)
4	1	3													(2)
4	7	7	9												(3)
5	2	4													(2)

(a) Write down the values needed to complete the stem-and-leaf diagram.
(b) Find the median and the quartiles of these times.
(c) Find the 67th percentile.
(d) On graph paper construct a box plot for these data, showing clearly your scale.
(e) Comment on the skewness of the distribution.

A random sample of trains arriving at the same main line station from Shefton had a minimum value of 15 minutes late and a maximum value of 30 minutes. The quartiles were 18, 22 and 27 minutes.

(f) On the same graph paper and using the same scale, construct a box plot for these data.

(g) Compare and contrast the train journeys from Darlingborough and Shefton based on these data

[ULEAC]

Revision

1   A police force publishes annual expenditure figures for salaries, vehicles, buildings and equipment. For a particular year, these are as follows, in millions of pounds:

salaries 3.7,   vehicles 1.2,   buildings 2.7,   equipment 1.4

Construct an accurate pie chart (radius 5 cm), showing clearly the angles for each sector. How much expenditure is represented by an angle of 1° on the chart?

2   The annual count of the number of visitors to a well-known stately home during the four quarters of a year is shown for two consecutive years.

	January–March	April–June	July–September	October–December
Year 1	15	25	35	15
Year 2	18	30	42	18

These figures are in thousands to the nearest thousand. Draw two accurate pie charts to illustrate these figures. If the radius of chart 2 is 4.8 cm, what must be the radius of the first, if the numbers are to be accurately compared?

3   A company manufactures three types of car, saloon, hatchback and estate, and each model is supplied with either a 1.6 litre or a 2.0 litre engine. The sales figures for the years 1985 and 1995 show the numbers sold by a franchise holder.

	1985		1995	
Engine size:	1.6	2.0	1.6	2.0
Saloon	17	26	25	13
Hatchback	23	20	27	14
Estate	19	15	24	12

Display these results on a compound bar chart so that the sales for the two years can be compared.

4   The figures below give the lives, in months to the nearest whole number, of 30 car batteries selected at random from the output of a manufacturer.

27   28   24   21   22   25   24   27   21   28   23   27   25   28   26
22   26   25   27   27   29   28   24   25   22   29   28   26   25   24

Draw a stem-and-leaf diagram to illustrate these figures, stating the (equal) class intervals used. If the last reading had been 30 instead of 24, how would you have divided up the figures so that the class intervals remained equal?

5   The temperature at noon on each day of a particularly warm and sunny June at a seaside resort was recorded as shown in the table. The figures represent degrees Celsius to the nearest 0.1 of a degree.

20.5   20.8   20.6   21.0   21.4   21.8   23.1   23.5   23.4   23.6
23.8   25.0   24.7   24.3   24.8   24.7   24.6   24.1   24.3   24.5
25.2   25.7   26.4   26.8   26.7   26.9   26.5   23.2   23.1   23.0

Using class intervals 20–20.9, 21–21.9, ... draw a stem-and-leaf diagram for these figures.

6   An author, encouraged by his long-suffering publisher, feels he must update his obsolete technology as he starts a new book. His expenditure for this is divided into four areas: (a) a new computer, (b) a printer, (c) furniture and books, (d) telephone calls for help! Given that he spent £175 on furniture and books, and the costs in (a) to (d) were in the ratio 40 : 12 : 7 : 1, respectively, calculate the sector angles for the chart and the costs of the remaining items.

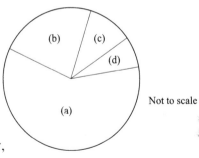

Not to scale

7   A class of 25 pupils takes tests in mathematics and physics (out of 100), with the following results:

Mathematics   33   34   38   40   42   42   45   47   49   49
              53   55   55   56   58   58   59   61   62   62
              66   72   73   76   84

Physics   28   30   31   33   33   37   39   41   43   44
              45   45   47   48   52   53   55   56   59   64
              69   73   78   82   85

Draw back-to-back stem-and-leaf diagrams, using class intervals 20–29, 30–39, ... for these figures and calculate the mean and standard deviation for each subject.

8 Choose your own class intervals to draw a stem-and-leaf diagram to illustrate the following measurements, in grams to the nearest gram, of 25 large randomly chosen Bramley cooking apples:

$$288 \quad 274 \quad 281 \quad 292 \quad 287 \quad 277 \quad 280 \quad 286 \quad 290 \quad 288$$
$$276 \quad 302 \quad 295 \quad 271 \quad 285 \quad 281 \quad 287 \quad 290 \quad 283 \quad 279$$
$$284 \quad 286 \quad 278 \quad 288 \quad 295$$

9 The masses of 20 male and 20 female pheasants, of equal ages, reared together on an estate farm and selected at random are given below. They are in grams to the nearest 10 g.

Males	1570	1630	1350	1600	1590	1670	1750	1480	1650	1410
	1680	1720	1560	1430	1730	1620	1690	1640	1680	1700

Females	980	1120	1030	870	1150	990	1150	1090	1010	1100
	810	1160	1090	790	1010	1050	1180	990	1050	1080

Draw a back-to-back stem-and-leaf diagram for these masses. Show that the average mass of a male pheasant is about 572 g more than that of a female.

10 On a single November day, the maximum daytime temperature in 20 UK towns and 20 towns outside Europe, chosen at random from records shown in a national newspaper, are displayed below. These figures represent degrees Celsius correct to one decimal place.

UK	12.2	11.1	7.2	11.1	7.8	7.2	12.2	10.0	2.8	10.0
	12.8	12.8	8.9	8.9	7.8	8.9	12.3	12.8	11.2	12.8

Abroad	17.2	23.9	10.0	12.8	31.1	30.0	22.2	22.2	17.2	11.1
	15.0	16.1	12.2	6.1	8.9	13.9	22.8	13.9	13.9	1.1

Display these figures on a back-to-back stem-and-leaf diagram.

11 A wine-taster visits all the supermarkets in her home town which have a wine department and tastes the red wines they sell. The numbers of wines in various price ranges are shown below, where W is the price (£). Draw a histogram to illustrate these figures.

W	$2.25 \le W < 3.25$	$3.25 \le W < 4.25$	$4.25 \le W < 5.25$	$5.25 \le W < 6.25$	$6.25 \le W < 7.25$
Number	6	34	46	10	4

*12 The 50 sumo wrestlers referred to in an earlier question have been in training after the tournament, and their masses, as the table shows, have increased. Masses M are given in kilograms to the nearest whole number.

M	$110 \le M < 125$	$125 \le M < 135$	$135 \le M < 145$	$145 \le M < 155$	$155 \le M < 175$
Frequency	2	13	22	9	4

(a) Calculate a new estimate for the mean mass.
(b) Construct a histogram to demonstrate this massive distribution!

2 | Probability

2.1 Basic laws, tree diagrams

Basic

1 A fair die is rolled once. Write down the probability that the score seen is:

(a) an odd number,   (b) greater than 4,   (c) an odd prime number.

2 A bag contains six red, three white and one blue disc. A single disc is removed at random. What is the probability that it is:

(a) not white,   (b) blue,   (c) white or blue,

(d) not red or blue?

3 The spinner shown is spun once and it is equally likely to stop anywhere round the circle. Write down the probability that the colour of the quadrant it stops at is

(a) red,
(b) green or yellow,
(c) not blue.

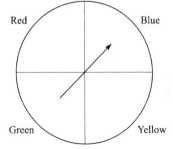

4 A school has five houses, A, B, C, D and E. A class in the school contains 23 pupils, four from house A, eight from B, five from C, two from D and the rest from E. A single student is selected at random to be the class representative on the school committee. What is the probability that the student is:

(a) from house B,   (b) not from houses D or E,
(c) from house A or house C?

*5 A bag contains 12 balls of which x are red, $2x$ are white and $3x$ are blue, where x is an integer. A single ball is selected at random. What is the probability that it is

(a) red,   (b) not white,   (c) white or blue?

6 Each square space on the dartboard shown
has the same area. A dart landing in an
unshaded square scores points as shown,
and a dart landing in a shaded square scores 4.
A single dart is thrown at random and hits the
board. What is the probability that the score is:

3	2	2	
2		3	2
2	3		2
	2	2	3

(a) 3,   (b) 2,   (c) less than 3,
(d) an odd number,   (e) 6,   (f) a multiple of 2?

7 A full pack of cards is laid face down on a table and a single card is
chosen at random. What is the probability that it is:

(a) an ace,   (b) a red ten,   (c) a five or a six,
(d) not a king or a queen?

8 Eight football teams reach the quarterfinals of a tournament. For the
draw, the teams are labelled A to H on eight discs placed in a bag, and
the discs are taken out, one at a time at random, and not replaced. What
are the following probabilities?

(a) The first disc chosen has a vowel on it.
(b) The second disc also shows a vowel.
(c) The third disc chosen is labelled G or H.

9 Cards are selected, one at a time, from a full pack without replacement.
Find the following probabilities:

(a) the first card is the ace of spaces,
(b) the second card is a red king assuming result (a),
(c) the third card is not a court card (i.e. a king, queen or jack),
      assuming results (a), (b).

10 A fair coin is spun twice. Draw a tree diagram to show all the outcomes
and their probabilities. Calculate the probability of seeing:

(a) no tails,   (b) at least one tail.

11 A fair die is rolled twice and the number of sixes seen is recorded.

(a) Draw a tree diagram showing all the possible outcomes and their
      probabilities.
(b) Find the probability that:

   (i) only one six is seen, (ii) fewer than two sixes are seen.

12 A bag contains three red and seven white discs. A single disc is chosen at
random, replaced, and then a second random selection is made.

(a) Draw a tree diagram to show all the possible outcomes.
(b) Write down the probability that:

   (i) only one red disc is seen,   (ii) at least one red disc is seen.

*13 The numbers 1 to 20 inclusive are written on small cards and put in a bag.

(a) A single card is chosen at random. Find the probability that the number on it is:

(i) even,   (ii) prime,   (iii) a perfect square.

(b) Two cards are removed at random, without replacement. Find the probability that the numbers on them:

(i) add up to 3,
(ii) are both perfect squares,
(iii) are both more than 15.

Intermediate

1 Two fair dice are rolled together once and the scores are added. Write out a table showing all the possible scores. What is the probability that the score is:

(a) more than 8,   (b) an even number,
(c) a prime number,   (d) not a multiple of 3?

2 Two fair dice are rolled together once and the differences (taken as positive) between the scores are recorded. Construct a table showing all the possible results. Write down the probability that the difference is:

(a) zero,   (b) an odd number,
(c) more than 3,   (d) an even prime number.

3 Two fair dice are rolled together once and the scores are multiplied. Write out the operation table for this event and find the following probabilities:

(a) the score is less than 5   (b) the score is a multiple of 9,
(c) the score is a perfect square

4 All the spades are removed from a full pack of cards. A card from this suit (i.e. spades) is chosen at random, is not replaced, and then a second random selection is made. Write down the following probabilities:

(a) the ace and king are selected in any order,
(b) the two and the three are chosen, in that order,
(c) no court card (king, queen, jack) is chosen.

*5 At a fête, a large piece of card has the numbers 1 to 1000 printed on it. Each player selects one number at random and that number cannot be reselected. The winning numbers are perfect squares greater than 500. What is the probability that

(a) the first player wins a prize.
(b) the second player fails to win, if the first has won,

(c) the third player wins a prize, assuming results (a), (b),
(d) the first player fails to win and the second does win a prize?

6 A burglary is known to have been carried out by two people. The police call in six men and four women for an identity parade. Two of the parade are randomly chosen by a witness. What is the probability that the two chosen are

(a) a man and a woman, in any order,
(b) two women,
(c) a woman and a man, in that order,
(d) at least one woman?

7 A fair coin is tossed three times. Draw a tree diagram to show all the possible outcomes. From this diagram, calculate the following probabilities:

(a) two heads are seen,
(b) one or two tails are seen,
(c) the result is tail, tail, head, in that order,
(d) more tails than heads are thrown.

8 Four fair coins are thrown together once. For each 'head' seen, 3 is scored, and for each 'tail' the score is 2. What is the probability that the score is:

(a) 8   (b) 11   (c) 9 or more   (d) 10 or 11?

9 Leela and Sachin sit two different examinations. You are given that p(Leela passes) $= \frac{1}{4}$, and p(Sachin passes) $= \frac{2}{5}$. Using the fact that these are independent events, find the probability that:

(a) both pass,   (b) only one passes,   (c) both fail.

10 A coin is biased so that p(heads) $= 0.7$. The coin is tossed three times. What is the probability that:

(a) one or two heads are seen,   (b) more tails than heads are seen?

11 A mathematics class contains ten boys and six girls. A team of three is to be chosen to enter a mathematical challenge competition. The selection is to be made at random. Draw a tree diagram to show the possible teams chosen. Find the probability that the team consists of:

(a) no boys,   (b) at least one girl,   (c) more girls than boys.

12 A fair coin is tossed three times. Use a tree diagram to calculate the following probabilities:

(a) the result is head, tail, tail, in that order,
(b) at least two heads are seen,
(c) either one or three tails are thrown.

13 The pupils in a school are divided into three political groups: Blues, Reds and Yellows. A mock election is held where all the candidates' names are written on the correct coloured discs and drawn at random from a bag, without being replaced. There are 200 Blues, 250 Reds and 150 Yellows. Write down the following probabilities:

(a) the first name drawn is a Blue,
(b) the second name is not a Red, given that the first was a Blue,
(c) the third name is a Red, given the first two results above.

14 A bag contains six apples, two of which are bad. Two apples are removed at random, without replacement. Use a tree diagram to calculate the following probabilities:

(a) both apples are bad,
(b) only one apple is good,
(c) the first apple is good and the second is bad.

15 A small unbiased cube has two of its faces painted red and the rest painted blue. The cube is rolled twice and the colour on its uppermost face is noted. Draw a tree diagram to show the possible outcomes. What is the probability that:

(a) a blue face is shown each time,
(b) one face of each colour is seen?

16 A biased die offers the following probabilities when it is rolled:

$$p(6) = \tfrac{1}{3}, \quad p(5) = \tfrac{1}{12}, \quad p(4) = \tfrac{1}{12}, \quad p(3) = \tfrac{1}{4}, \quad p(2) = \tfrac{1}{16}, \quad p(1) = \tfrac{3}{16}$$

The die is rolled once. Write down the probabilities of the following scores:

(a) 6 or 1,
(b) an even number,
(c) an even prime number,
(d) a multiple of 3,
(e) less than 6.

A second die is biased so that $p(6) = \tfrac{1}{5}$, the other scores being equally likely. The two dice are rolled together once. Find the probability that the total score is (f) 12, (g) 10, (h) 3.

Advanced

1 A set of playing cards consists of all the aces, kings, queens and jacks from a complete pack. From these sixteen cards, two random selections are made, without replacement. Draw a tree diagram illustrating these events and use it to find the probability that:

(a) the two cards are alike (e.g. two aces),
(b) one ace is chosen,
(c) no queens are seen,
(d) two red jacks are chosen,
(e) a red king and a black queen are selected.

2   A small, unbiased cube has three faces painted green, two painted blue and the last face painted yellow. The cube is rolled twice and the colour of the uppermost face is recorded. Draw a tree diagram to show all the possible outcomes and the probability of each result. What is the probability that:

(a) the colours seen are the same,
(b) a yellow face is seen on the second roll but not the first,
(c) no green face is seen,
(d) a green and a yellow face appear?

*3   Harry is going on holiday to a tropical island and he has to be immunised against the lethal purple bug disease. The probability that his injection is successful is $\frac{4}{5}$. If it is successful then the probability that he catches the disease is $\frac{1}{5}$, but if it is not successful then the probability that he catches the disease goes up to $\frac{3}{5}$. If he catches the disease, he is given treatment which has a 75% probability of success. Draw a tree diagram to illustrate all these events, and from it calculate the following probabilities:

(a) the immunisation fails, but he doesn't catch the disease,
(b) the immunisation is successful, but he catches the disease, and the subsequent treatment fails,
(c) he needs treatment for the disease.

4   A biased die and a biased coin are such that $p(6) = \frac{1}{3}$ and $p(\text{head}) = \frac{1}{4}$ respectively. The die is rolled twice and then the coin is spun once. Draw a tree diagram, starting with the two rolls of the die, to show all the possible outcomes of these events. Find the probability that the result is:

(a) no sixes are seen,
(b) a tail is thrown,
(c) one throw of the die shows a six and the other does not,
(d) at least one six is thrown,
(e) one six and one head are seen.

5   Of the pumpkins grown on allotments in the village of Greenville, 60% are fertilised with Vegswell, a popular organic fertiliser. If this fertiliser is used, then the probability of a pumpkin exceeding 3 kg when picked (they are all picked after the same growing time) is 0.7, but, without Vegswell, the probability drops to 0.2. Draw a tree diagram to show all the

outcomes and probabilities for these events. Find the probability that, when picked,

(a) a fertilised pumpkin weighs less than 3 kg,
(b) an unfertilised pumpkin weighs more than 3 kg,
(c) a pumpkin chosen at random weighs more than 3 kg.

6 Simon takes two successive tests in mathematics and the probability that he passes the first is $\frac{3}{5}$. If he passes the first, the probability that he passes the second is $\frac{4}{5}$ but, if he fails the first, then this probability goes down to $\frac{2}{5}$. Draw a tree diagram to illustrate these events and from it calculate the probability that:

(a) he passes both examinations at the first attempt,
(b) he fails the first but passes the second,
(c) he passes one of the tests.

7 A dartboard consists of three concentric circles as shown. The scores when the dart lands in the three areas A, B, and C are 2, 5 and 10 respectively. The radii of the circles are R_1 cm, R_2 cm and R_3 cm as in the diagram. A single dart is thrown at random and it hits the board.

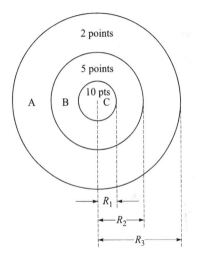

(a) Calculate the probability, in terms of R_1, R_2 and R_3, that the score is (i) 2, (ii) 5, (iii) an even number.
(b) If two darts are thrown at random, both hitting the board, what is the probability that the total score is an odd number?

8 A multiple-choice examination question offers four possible answers, one of which is correct. The probability of a student guessing the correct answer is p and, if an incorrect answer is chosen, each is equally likely to be selected. Two students, neither knowing the correct answer, each make a guess. What is the probability that:

(a) both select the correct answer,
(b) one chooses correctly and the other wrongly,
(c) each selects a different wrong answer?

9 In a school, house A has 20 boys and 15 girls, and house B has 15 boys and 30 girls. A single member of house A, chosen at random, is transferred to house B. Find the probability that, after this transfer:

(a) if one member of house A is selected at random, it is a boy,
(b) if one member of house B is selected at random, it is a girl.

10 A bag contains three red beads, three blue beads and three green beads, which are indistinguishable apart from colour. The beads are thoroughly mixed. One bead is removed and placed on a table; a second bead is then removed.

(a) By using a tree diagram, or otherwise, find the probability that the two beads are the same colour.

(b) Another bag contains three red beads, three blue beads and two green beads. Two beads are removed as before. Show that the probability that the two beads are the same colour is the same as in part (a).

Revision

1 A tetrahedral die has its faces marked 1, 3, 5 and 7. It is biased so that $p(1) = \frac{1}{3}$, the other scores being equally likely. The die is rolled twice; find the probability that the total of the scores is:

(a) 4,
(b) 8,
(c) 14,
(d) an odd number,
(e) more than 8, if the first score was 3.

2 The twos, threes, fours, and fives are removed from a full pack of cards. From these 16 cards, four are chosen at random and without replacement. Find the probability that they are:

(a) all red,
(b) all spades,
(c) all fives,
(d) two threes and two fours.

3 A pair of cubical dice, D_1 and D_2, are rolled together once. D_1 is biased so that $p(1) = \frac{1}{4}$, the remaining probabilities being equal to each other. D_2 is biased according to the probability distribution below:

Score	1	2	3	4	5	6
p(that score)	$\frac{1}{10}$	$\frac{1}{5}$	$\frac{1}{10}$	$\frac{1}{5}$	$\frac{1}{10}$	$\frac{3}{10}$

What is the probability that the scores on D_1 and D_2 are as listed:

	D_1	D_2
(a)	3	4
(b)	6	6
(c)	even number	odd number?

(d) What is the probability that the total of the scores is 4?

4   The arrow shown, when spun, is equally likely to stop at any point around the circle. The numbers in the sectors are the scores achieved if the arrow stops over that sector. If the arrow stops exactly over the dividing radius between two sectors, it is spun again. The arrow is spun once.

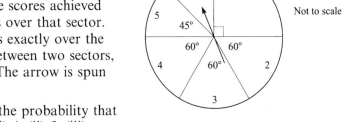

Not to scale

(a) Write down the probability that the score is (i) 4, (ii) 5, (iii) an odd number.

The arrow is now spun twice and the scores are added.

(b) What is the probability that the total score is (i) 2, (ii) 11, (iii) 9, (iv) 7?

5   Three cubical dice are each biased so that $p(6) = \frac{1}{2}$, the other scores being equally likely. The dice are rolled together once. What is the probability that the total score is:

(a) 16, if one die shows a 4,
(b) 16, if one die shows a 5,
(c) 16, if one die shows a 6?

6   The eleven letters of the word PHILIPPINES are written on small cards and put in a bag. They are taken out one at a time, at random, without replacement. Find the following probabilities:

(a) the first three letters, in order, spell the word SIP,
(b) the first four letters, in order, spell the word PINE,
(c) the first five letters, in order, spell the word SNIPE,
(d) the first six letters, in order, spell the word PHILIP.

7   A circular dartboard has four sectors, with angles of 30°, 60°, 120° and 150° as shown. The scores made when a dart lands in a sector, 4, 3, 2 and 1, are also shown.

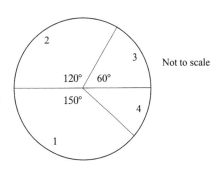

Not to scale

(a) A single dart is thrown at random and hits the board. What is the probability that the score is (i) 1, (ii) an even number, (iii) a prime number?

(b) Two darts are thrown at random and both hit the board. Find the probability that:

    (i) they both land in the largest sector,
    (ii) they give a total score of more than 5,
    (iii) they give a total score of 3.

8 Nadeem sits three examinations. The probability that he passes the first is $\frac{3}{4}$. The probability that he passes the second is $\frac{2}{3}$ if he passes the first, but $\frac{1}{2}$ if he fails the first. The probability that he passes the third is $\frac{3}{5}$ if he passes the second, but $\frac{1}{5}$ if he fails the second. Draw a tree diagram to show these events and probabilities. Find the probability that:

(a) he passes only the first examination,
(b) he fails all three examinations,
(c) he passes two of the examinations.

9 The members of a club decide to hold a sweepstake to raise funds. Two races are selected, the first having 12 horses entered and the second 10 horses. Prizes will be won if:

    (i) the first two horses are correctly guessed in the right order in the first race,
    (ii) the first three horses in the second race are guessed in the right order,
    (iii) both winners are chosen.

What are the probabilities of success in the following cases:

(a) John chooses two horses in the first race,
(b) Shirley chooses three horses in the second race,
(c) Ranjit selects, he hopes, two winners?

10 There are 40 students taking A level examinations in mathematics, physics and chemistry. Some take all three subjects, some take two and some study only one of the three. The totals taking mathematics, physics and chemistry are 22, 16 and 18 respectively. You are also given that five take mathematics and physics only, four study mathematics and chemistry but not physics, and three study physics and chemistry, but, controversially, not mathematics. Six study only physics. A single student is selected at random. Find the probability that she is studying

(a) only mathematics,   (b) all three subjects.

*11 A die and a coin are both biased. For the die, $p(6) = \frac{3}{5}$, the other scores being equally likely. For the coin, $p(\text{tails}) = \frac{7}{10}$. The die is rolled once and the coin is spun once.

(a) Find the probabilities of the following results:

    (i) (six, heads),   (ii) (three or four, tails).

(b) In a second experiment, the die is rolled twice, the total score being recorded, and the coin is again spun once. Find the probability that the total score is 11, while the coin shows tails.

12 An unbiased cubical die has the letters A to F on its six faces.

(a) The die is rolled once. What is the probability of seeing a vowel?
(b) The die is now rolled twice. Find the probabilities of the following results:

   (i) a vowel is seen both times,
   (ii) a vowel and a consonant are seen.

(c) The die is now rolled four times. What is the probability that the four results are:

   (i) the letters F, A, C, E, in that order,
   (ii) two vowels and two consonants?

13 A bag contains six dark blue, nine light blue and five purple discs. Random selections of single discs are made without replacement. Find the following probabilities:

(a) the first three discs chosen are dark blue,
(b) neither of the first two discs chosen is purple,
(c) one disc of each colour is chosen.

14 A dart is thrown at the board shown and hits the sector marked 8 with probability 0.6. How many darts must be thrown at the board if the probability of scoring at least one 8 is at least 0.95? Assume every dart hits the board.

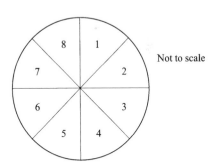

Not to scale

2.2   Conditional probability, arrangements and selections

Basic

1 A and B are events such that p(A) = 0.4, p(B) = 0.3 and p(A ∪ B) = 0.5. Calculate:

(a) p(A ∩ B)   (b) p(A | B)   (c) p(B | A)
(d) p(A')   (e) p(B' ∩ A)

2 The figures in the Venn diagram indicate the number of elements in the sets. Write down:

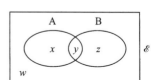

(a) n(A ∪ B)   (b) n(A ∩ B′)
(c) p(A | B)   (d) p(B | A′)
(e) n(A ∪ B)′

3 (a) In the Venn diagram shown, find the values of *w, x, y* and *z,* where these are the numbers of elements in the sets, given the following information:

$$n(\mathscr{E}) = 20, \ n(B) = 10, \ p(A \mid B) = 0.4,$$
$$n(A) = 5.$$

(b) Write down the following probabilities:

(i) p(B) (ii) p(A ∪ B)′ (iii) p(B | A′)

4 You are given that A and B are events such that

$$p(A \mid B) = \tfrac{1}{2}, \quad p(B) = \tfrac{3}{5}, \quad p(A \cup B) = \tfrac{4}{5}.$$

Find

(a) p(A ∩ B),   (b) p(A),   (c) p(B | A′),   (d) p(A ∪ B)′,   (e) p(A′ ∪ B).

*5 Ten cards, labelled A, B, C, D, E and 1, 2, 3, 4, 5 respectively, are mixed and placed face down on a table. Two cards are selected, at random, with replacement. Find the following probabilities:

(a) the same vowel is drawn twice,
(b) two numbers, which add up to 4, are drawn,
(c) given that two numbers are drawn, they add up to 10.

6 Eight cards are labelled as shown:

| F | A | S | T | | 1 | 2 | 3 | 4 |

The two groups, separate from each other, are mixed, placed face down on a table, and one card from each group is selected at random.

(a) Find the following probabilities:

(i) the cards A, 1 are selected, in that order,
(ii) card F is selected first and then card 2 or card 3,
(iii) neither card T nor card 4 is selected.

(b) The cards are now mixed and placed face down on the table, and two are randomly chosen without replacement. Give that both are numbered cards, what is the probability that the sum of the numbers is 5?

7 (a) A fair die is rolled once. Given that the score is an even number, find the probability that it is (i) a prime number, (ii) not a multiple of three.

(b) If the die were rolled twice, what is the probability that the total of the two scores is an even number?

8 The spinner shown is equally likely to stop at any point round the circle. If it stops over a radius, it is spun again. If it is spun once, find the following probabilities:

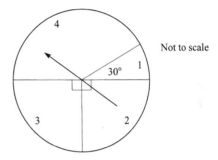

Not to scale

(a) the score is 3,
(b) the score is a prime number,
(c) the score is 3, given that it is a prime number,
(d) the score is 4, given that it is an even number.

9 Eight athletes contest the final of a 400 m race. The first three win a medal. In how many different ways can the three medals be awarded?

10 A committee of three is to be chosen from eight students. How many different committees can be chosen? If two of the students are sisters, how many of the committees will contain both sisters?

11 Six people sit at a round dinner table. How many different seating arrangements can be made (there are only six seats round the table) if clockwise and anti-clockwise arrangements are regarded as (a) the same, (b) different?

12 A committee is to consist of two boys and three girls, to be selected from five boys and eight girls. How many different committees can be chosen?

Intermediate

1 A football team is made up of two strikers, four midfield players, four defenders and a goalkeeper. Before an important match, the eleven players stand in a line facing the main stand, to be greeted by a prominent visitor. If the four defenders, labelled D_1, D_2, D_3, D_4, always stand next to each other, in that order from left to right, how many different line-ups are possible?

A short-sighted fan, who has forgotten her glasses and cannot distinguish any player, says that the fourth player from the left is the best player in the team. What is the probability that he is a defender?

2 The members of a mixed rowing eight are to be selected from a group of nine men and five women. How many selections can be made, taking no account of different positions in the boat, if (a) only men are chosen, (b) two women are selected, (c) at least three women are chosen, (d) all the women are chosen and they have to sit together in the boat?

3 Use your calculator to find the value of:

(a) $6! + 4!$ (b) $5! - 3!$ (c) $\dfrac{6!}{4!}$

(d) $\dfrac{8!}{4!3!}$ (e) $\dfrac{6!}{3!2!}$ (f) $\dfrac{5!}{2! + 4!}$

(g) $\dfrac{10!}{5! + 6!}$ (h) $\dfrac{7!}{4! - 3!}$ (i) $5!(3! + 4!)$

(j) $\dfrac{3! + 4!}{2! + 3!}$ (k) $\dfrac{5! - 3!}{3!}$ (l) $\dfrac{(x+2)!}{(x-1)!}$

(m) $\dfrac{(x-3)!}{(x-2)!}$ (n) $\dfrac{(x+2)! + (x+1)!}{(x+2)!}$ (o) $\dfrac{x!}{x! + (x+1)!}$

4 In how many ways can the letters of the word POTATO be arranged, if all are used each time? How many of these arrangements (a) begin with A, (b) do not begin or end with P, (c) have the two Ts together?

5 A man attempts to predict the results of five football matches from a set of ten. How many answers can be made if the possible results are 'home win', 'draw', 'away win'?

6 Six boys and two girls are placed in a line. What is the probability that the girls are separated?

*7 A bag contains ten coloured discs. Three are yellow, five are green and two are brown. They are numbered as follows: Y_1, Y_2, Y_3, $G_1, \ldots, G_5$, B_1, B_2. Two discs are removed at random, without replacement. Find the probability that (a) they are G_1, B_1, in any order, (b) both are green, (c) one is brown, given that the other is G_5.

8 A box contains six plain, nine milk and five white chocolates. Three are selected, without replacement. Find the probability that the set selected contains (a) three plain chocolates, (b) one plain and two milk chocolates, (c) no white chocolates, (d) one of each kind of chocolate.

9 A fair die is rolled twice. Calculate the probability that:

(a) given that the first score is five, the total score is an even number,

(b) the total score is six, given that the first number is odd,

(c) the total score is a prime number, given that the first number is odd.

10 From a full pack of playing cards, two are removed at random (i) without replacement, (ii) with replacement. In each case find the probability that:

(a) the second card is the same as the first,

(b) a red king and a black jack are drawn,

(c) the second card is red, given that the first is black,

(d) the second card is an ace, given that the first is not an ace,

(e) both cards are aces.

11 A university student is trying to make up his mind which of his textbooks he should take with him on a revision weekend. He is studying history and French and has six history books, $H_1, \ldots, H_6$, and five French texts, $F_1, \ldots, F_5$. As there is only a limited time, he takes only five books, and, being undecided, selects them at random. Find the probability that his combination is:

(a) all five books are on history,

(b) H_1, H_2, H_3 are taken, with F_1 and F_2,

(c) three history books and two French books,

(d) only one French text and four history books.

*12 A GCSE mathematics class consists of 12 boys and 16 girls.

(a) A single pupil is selected at random. Find the probability that he is a boy.

(b) Two pupils are selected at random and without replacement. Find the probability that (i) both are girls, (ii) one is a boy and one is a girl.

(c) Three pupils are selected at random, with replacement. What is the probability that:

(i) they are boy, girl, boy, in that order,

(ii) all three are girls,

(iii) more girls than boys are chosen?

13 A litter of eight puppies is born to a black labrador bitch. All are black except that three have a white spot on their backs and two others have a white ring around their tails. There are no other distinctive markings.

(a) A single pup is chosen at random. What is the probability that it will not have a white spot?

(b) Two pups are to go to a particular family. If they are selected at random, what is the probability that (i) both are plain black, (ii) one has a spot and the other a ring?

(c) If the two pups in (b)(ii) have already been taken away and three of those remaining are to be selected at random, what is the probability that (i) none is plain black, (ii) one has a spot?

14 A different whole number is written on each of fifteen cards. Ten of these are even, five are odd. A single card is chosen at random. If the card shows an even number, the player tosses a fair die, but if the number is odd, the player must toss a biased die. This die has $p(6) = \frac{1}{4}$, the other scores being equally likely.

(a) Find the probabilities of the following scores: (i) even number, 6, (ii) odd number, 1.
(b) Given that the number on the card was odd, what is the probability that the die tossed showed a prime number?

15 Each section of the tree diagram shows the results of tossing coins. All but one of the coins is biased as shown, and the player has to follow the order given, i.e. if he throws a head with his first coin, he has to toss the coin for which $p(\text{heads}) = \frac{3}{4}$, and so on. After the three sections are completed, find the probability that:

(a) two heads were seen,
(b) the middle result was a head,
(c) more tails than heads were seen.

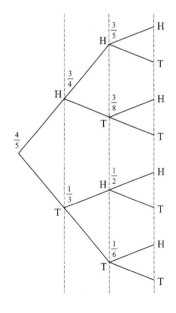

Advanced

1 A security lock on a door has four buttons, which must be pressed in a particular order to open the lock. Someone in a hurry has forgotten the code and punches buttons at random. How many completely wrong attempts can be made (i.e. no button is pressed in its correct position)? Repetition is allowed.

2 You are given that x and y are positive integers, defined as follows:

$$1 \le x \le 5, \qquad 2 \le y \le 6.$$

One number from each set is selected at random. Find the probability that:

(a) the sum of the numbers is an odd number,
(b) the sum is 5, given that it is a prime number,
(c) the number from set x is even, given that the sum of the scores is odd,
(d) the x-number is odd and the y-number is even.

3 A student is offered three courses, A, B, and C, from which one must be chosen, for his first year of A level studies. For the second year, there are two choices, P and Q. The head of sixth form studies knows from past experience that the proportions of students making the choices are A 40%, B 50% and C 10% while, for year 2, 70% choose course P. Find the probabilities of the following choices, based on the given proportions:

(a) the student chooses course A and Q,
(b) given that course C was not chosen, course P is selected in the second year,
(c) course Q is chosen for year 2 and course B was not selected for year 1.

*4 A seed merchant sells three types of flower seed, F_1, F_2 and F_3. They are sold, by public demand, as a mixture where the proportions are $4:4:2$ respectively. The germination rates of the three types are 45%, 60% and 35% respectively. Draw a tree diagram to illustrate this data. From it calculate the probability that:

(a) a randomly chosen seed will germinate,
(b) given that the seed is of type F_3, it will not germinate,
(c) given that a randomly chosen seed does not germinate, it is of type F_2.

5 A bag contains three red, five white and two blue discs. Two discs are removed at random, without replacement.

(a) Draw a tree diagram to illustrate these events.

Calculate the probability that the two discs are:
    (i) both blue, (ii) of the same colour, (iii) one red and one blue.

(b) The first two discs are kept out of the bag and a third disc is selected at random

    (i) Given the first two discs are white, what is the probability that the third disc is not white?
    (ii) Given that a red and a white disc were removed, what is the probability that the third disc is not blue?

6 A football fan is taking part in a sports quiz. He is being asked (i) the name of particular team, (ii) the results of particular matches played by that team. The following probabilities apply:

p(he remembers the team) $= 0.6$,   p(he forgets the match result) $= 0.45$.

Calculate the probability that:

(a) he gives both answers incorrectly,
(b) given he names the team correctly, he gives the correct result.
(c) he names the team wrongly, but gives the correct result.

7 A school places its students into four houses A, B, C and D. The cricket coach selects a squad of 15 players from whom he will select the first eleven.

(a) Given that the squad is made up of three members of house A, five from B, one from C and six from D, find the probability that the final eleven, if chosen at random (because all fifteen are all-rounders), contains

   (i)   all the five players from house B,
   (ii) two players from house B,
   (iii) the one player from house C.

(b) How many different teams can be selected irrespective of house or ability of the players?

8 A die is biased so that the following probabilities apply:

Score	1	2	3	4	5	6
p(that score)	0.2	0.1	0.15	0.2	0.2	0.15

(a) The die is rolled once. Find the probability that:

   (i)   the score is an even number,
   (ii) the score is a prime number less than 5,
   (iii) the score is three or five?

(b) The die is now rolled twice and the scores added. What is now the probability that:

   (i)   given the score is an odd number, it is a prime number,
   (ii) given the score is less than 7, it is an odd number,
   (iii) given the score is an even number, it is a multiple of three?

9 In a competition a long-jumper takes three jumps. The probability that her second jump is longer than her first jump is 0.6. If her second jump is longer than her first then the probability that her third jump is the longest of the three is 0.2. If her first jump is longer than her second jump then the probability that her third jump is the longest of the three is 0.3. The information concerning the results of her second and third jumps is shown in the following tree diagram.

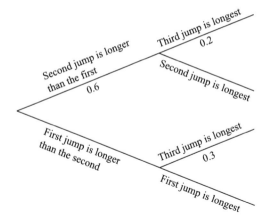

(i) Find the probability that, after her first jump, she improves with each subsequent jump.
(ii) Find the probability that she improves on her first jump.
(iii) Find the conditional probability that her second jump is longest, given that she improves on her first jump.

[UCLES]

10 *Alpha Bet* is a weekly lottery in which a selection of 5 *different* letters, chosen at random from the 26 letters A to Z inclusive, appear on each ticket. Each ticket has a different combination of letters printed on it. At the end of a week, 5 winning letters are chosen. Each person buying a ticket for the lottery pays £1.
Prizes are awarded depending on the number of letters on the ticket which match with the winning letters. For example, when the winning letters are D, G, K, M, X, a ticket with the letters E, K, M, R, X, has three matching letters. The order of the letters on a ticket does not matter, nor does the order in which they are drawn.
Values of prizes are:

Number of matching letters	3	4	5
Prize	£10	£100	£1000

(i) What is the largest number of tickets that could be sold in a week?

Consider a week when all possible tickets are sold.

(ii) What is the probability that I win the £1000 prize, if I buy only one ticket?
(iii) If the winning letters are D, G, K, M, X, give *two* examples of different tickets which would win a prize of £100. Show that altogether there are 105 tickets which would win a prize of £100.
(iv) Find the number of tickets that would win a prize of £10.

(v) Show that the chance of a ticket winning a prize is roughly 1 in 30.

(vi) Show that about half the stake money is returned in prizes.

[OCSEB/UCLES]

Revision

1 A poker hand of five cards is dealt at random from a full pack. What is the probability that it consists of three aces and two kings?

2 In how many ways can the letters of the following words be arranged, if all are used every time?

(a) MACHINE (b) STATISTICS (c) EXAMPLE

(d) TESSELLATE (e) TERRIER

3 Ten horses are entered for a five-furlong sprint race and, if a race-goer chooses the first three in the correct order, the prize is a large sum of money. What is the probability that he makes the correct choice?

4 I_1, I_2, I_3, I_4 and I_5 are five different positive integers between 0 and 9 inclusive, such that $I_1 < I_2 < I_3 < I_4 < I_5$. How many different numbers can be made, using all five integers, if

(a) I_1 and I_5 are the first and last digits respectively,

(b) I_2 and I_4 are always next to each other

5 If $^nC_2 = 15$, calculate the value of n.

6 A bag contains 12 red and 8 white cards.

(a) A single card is removed at random.

Find the probability that it is (i) white, (ii) red or white.

(b) A second experiment consists of removing two cards in succession. Assuming (A) no replacement, (B) replacement, find the probability that:

(i) both cards are red,

(ii) one card of each colour is seen,

(iii) the second card is white given that the first was red.

7 A and B are sub-sets of a universal set $\mathcal{E}$, such that $n(A \cap B) > 0$, $n(\mathcal{E}) = 20$. You are given that $n(A) = 7$, $n(B) = 9$, $n(A' \cap B) = 7$. A single element of this set is chosen at random. What is the probability that it is an element of:

(a) $A \cap B$ (b) A' (c) $(A \cup B)'$ (d) $A \mid B'$?

8 Ten cards, labelled A, A, A, B, B, D, E, E, T, T are shuffled and put face down on a table. They are selected, one at a time, at random, without replacement. Show that it is equally likely to see the letters BAT or TEA and also the words DATE or BEAD in those orders.

*9 From a full pack of cards, five cards are dealt out at random, without replacement. Show that the probability of seeing three aces and two kings, in any order, is 24 times the probability of seeing the ace, king, queen, jack and ten of spades, also in any order.

10 A committee of three is to be chosen from a group of five men and three women. Assuming random selection, what is the probability that the committee will consist of (a) three men, (b) three women, (c) two men and one woman, (d) more men than women?

11 Two red discs and eight white discs are to be arranged in a straight line. How many different arrangements are there if:
(a) the two red discs are together at either end of the line,
(b) the two red discs are always next to each other?

What would the answers to (a) and (b) be if the red discs were numbered R_1 and R_2 and the white discs numbered $W_1, W_2, \ldots, W_8$?

12 (a) In how many ways can the letters of the following words be arranged, if all the letters are used every time:
(i) CHURCH   (ii) RADAR   (iii) CALCULATOR   (iv) ISOSCELES
(v) STRIKERS
(b) How many different groups of three letters can be formed from the letters of the word PIPPIN?

3
Discrete distributions

3.1 Probability distribution functions, expectation and variance

$$E(X) = \sum xp, \quad Var(X) = \sum x^2p - \{E(X)\}^2$$

Basic

Abbreviations used:

pdf probability distribution function
drv discrete random variable

1 In each line of the table below, the pdf of a drv is defined. Find the value of k for each definition.

Values of x	$p(X = x)$
(a) 1, 2, 3	$k(4 + x)$
(b) 3, 4, 5	$\dfrac{kx}{12}$
(c) 1, 2, 3	$k(x^2 - 2)$
(d) 2, 3, 4	$\dfrac{kx}{27}$
(e) $\frac{1}{2}, \frac{1}{3}, \frac{1}{4}$	$\dfrac{kx}{2}$
(f) 1, 2, 3, 4	$k(2x - 3)$
(g) 0, 1, 2	$k(\frac{1}{2})^x$
(h) 2, 3, 4, 5	$\dfrac{2x + 1}{k}$

2 Using the definition of a pdf, show that the variable X is a drv for the values of x given.

Values of x	$p(X = x)$
(a) 10, 11, 12	$\frac{1}{39}(x + 2)$
(b) 4, 5, 6, 7	$\frac{1}{10}(x - 3)$
(c) 0, 1, 2, 3	$\frac{1}{18}(x^2 + 1)$
(d) 2, 3, 4	$\frac{1}{23}(x^2 - 2)$
(e) 3, 4, 5	$\dfrac{60}{47x}$
(f) 2, 3, 4	$\dfrac{315}{401}\left(\dfrac{x}{2x + 1}\right)$
(g) 0, 1, 2, 3, 4	$\frac{1}{10}(x^2 - 5x + 6)$
(h) 0, 1, 2, 3	$\frac{1}{76}(x + 1)^x$

3 Five examples of pdf's for the drv X are shown below. Find a formula for each pdf.

(a)

x	0	1	2	3
$p(X = x)$	$\frac{1}{10}$	$\frac{2}{10}$	$\frac{3}{10}$	$\frac{4}{10}$

(b)

x	2	3	4	5
$p(X = x)$	$\frac{3}{24}$	$\frac{5}{24}$	$\frac{7}{24}$	$\frac{9}{24}$

(c)

x	10	11	12	13
$p(X = x)$	$\frac{4}{22}$	$\frac{5}{22}$	$\frac{6}{22}$	$\frac{7}{22}$

(d)

x	2	3	4	5
$p(X = x)$	$\frac{1}{30}$	$\frac{4}{30}$	$\frac{9}{30}$	$\frac{16}{30}$

(e)

x	0.01	0.02	0.03	0.04
$p(X = x)$	0.1	0.2	0.3	0.4

4 A bag contains 12 green and 8 blue discs. A disc is removed at random and then replaced. This is done three times. If X is the drv representing 'the number of green discs seen', write out the probability distribution of X.

5 The arrow in the spinner shown is equally likely to land on any sector. It is spun three times. Write out the pdf for the drv X, 'the number of times the spinner lands on the sector numbered 3'.

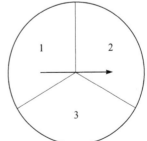

6 Two fair dice are rolled together once. The drv X is defined as 'the number of sixes seen'. Write down the pdf of X.

7 For the following pdf's, find $E(X)$ and $E(X^2)$.

(a)

x	1	2	3	4
$p(X = x)$	0.1	0.2	0.3	0.4

(b)

x	5	6	7	8
$p(X = x)$	0.05	0.3	0.4	0.25

(c)

x	-4	-3	-2	-1	0
$p(X = x)$	0.1	0.2	0.4	0.2	0.1

(d)

x	2	3	4	5
$p(X = x)$	$\frac{1}{5}$	$\frac{2}{5}$	$\frac{1}{5}$	$\frac{1}{5}$

(e)

x	10	11	12	13
$p(X = x)$	0.04	0.34	0.54	0.08

(f)

x	-1	0	1	2	3	4
$p(X = x)$	0.01	0.15	0.25	0.37	0.2	0.02

(g)

x	25	26	27	28
$p(X = x)$	$\frac{1}{10}$	$\frac{3}{10}$	$\frac{4}{10}$	$\frac{2}{10}$

(h)

x	-2	-1	0	1	2	3
$p(X = x)$	0.1	0.28	0.31	0.17	0.09	0.05

(i)

x	3	4	5	6	7
$p(X = x)$	$\frac{1}{8}$	$\frac{2}{8}$	$\frac{3}{8}$	$\frac{1}{8}$	$\frac{1}{8}$

(j)

x	20	21	22	23	24
$p(X = x)$	$\frac{1}{25}$	$\frac{4}{25}$	$\frac{11}{25}$	$\frac{6}{25}$	$\frac{3}{25}$

8 In the five parts of this question, the drv X has a pdf defined. Calculate the quantities requested for each definition.

(a)

x	2	3	4
$p(X = x)$	0.2	0.5	0.3

Calculate (i) $E(X)$, (ii) $E(4X)$, (iii) $E(4X - 3)$, (iv) $E(X^2)$, (v) $E(2X^2 - 1)$.

(b)

x	0.1	0.2	0.3	0.4
$p(X = x)$	0.1	0.3	0.4	0.2

Calculate (i) $E(X)$, (ii) $E(\frac{1}{2}X)$, (iii) $E(\sqrt{X})$, (iv) $E(\sqrt{X} + 2)$, (v) $E(X^2)$.

(c)

x	1	2	3	4	5
$p(X = x)$	$\frac{1}{10}$	$\frac{2}{10}$	$\frac{3}{10}$	$\frac{3}{10}$	$\frac{1}{10}$

Calculate (i) $E(X)$, (ii) $E(10X)$, (iii) $E(10X - 5)$, (iv) $E(3X^2 - 10)$.

(d)

x	-2	-1	0	1
$p(X = x)$	0.4	0.3	0.2	0.1

Calculate (i) $E(X)$, (ii) $E(2X)$, (iii) $E(5 - 2X)$, (iv) $E(X^2 + 1)$.

(e)

x	-4	-3	-2	-1
$p(X = x)$	0.1	0.4	0.3	0.2

Calculate (i) $E(X)$, (ii) $E(2X^2)$, (iii) $E(X^3)$.

9 Find the expected number of heads seen when
(a) two fair coins are tossed together once,
(b) a single fair coin is tossed three times.

*10 The pdf's for several drv's are shown below. Calculate the required quantities for each variable.

(a)

x	1	2	3	4
$p(X = x)$	0.05	0.35	0.4	0.2

Calculate (i) $E(X)$, (ii) $E(X^2)$, (iii) $\mathrm{Var}(X)$, (iv) $\mathrm{Var}(2X + 3)$.

*(b)

x	2	3	4	5
$p(X = x)$	$\frac{1}{4}$	$\frac{3}{8}$	$\frac{1}{4}$	$\frac{1}{8}$

Calculate (i) $E(X)$, (ii) $E(X^2)$, (iii) $\mathrm{Var}(X)$, (iv) $\mathrm{Var}(3X - 1)$.

(c)

x	-2	-1	0	1
$p(X = x)$	0.4	0.25	0.2	0.15

Calculate (i) $E(X)$, (ii) $E(2X^2 + 1)$, (iii) $\mathrm{Var}(X)$, (iv) the standard deviation of X.

(d)

x	0.1	0.2	0.3	0.4
$p(X = x)$	0.15	0.25	0.35	0.25

Calculate (i) $E(X)$, (ii) $E(3X^2)$, (iii) $\mathrm{Var}(2X)$, (iv) the standard deviation of X.

(e)

x	0	1	2	3	4
$p(X = x)$	0.1	0.25	0.3	0.2	0.15

Calculate (i) $E(X)$, (ii) $E(2X^2 - 3)$, (iii) $\mathrm{Var}(\frac{1}{2}X)$, (iv) the standard deviation of X.

11 The pdf shown defines a drv X where $E(X) = 2.7$. Show that $a = 3$.

x	1	2	a	4
$p(X = x)$	0.1	0.3	0.4	0.2

12 The pdf of the drv X is shown below. You are given that $b > 0$ and that $E(X^2) = 20$. Show that $b = 6$.

x	2	4	b	8
$p(X = x)$	0.4	0.3	0.2	0.1

13 Find the expectation and variance of the drv X in each of the following.

(a) A fair die is rolled three times. X is 'the number of "ones" seen'.

(b) A biased die is such that $p(4) = 0.1$, all the other scores being equally likely. The die is rolled twice. X is 'the number of fours seen'.

Intermediate

1 Find the expectation and variance of the drv X in each of the following. Write down the pdf in each case.

(a) Jafar has a fair coin and Jameela a fair die. Jafar first tosses the coin. If it shows heads, Jameela rolls the die once; if it shows tails, she rolls twice. X is the drv 'the number of sixes seen'.

(b) There are five children in a family of which three are boys.

(i) Two children are selected at random. X is the drv 'the number of boys chosen'.

(ii) Three children are chosen at random. Again, X is the drv 'the number of boys chosen'. Both selections are without replacement.

(c) Two biased dice are thrown together once. For the first, $p(6) = 0.5$, the other scores being equally likely. For the second, $p(1) = 0.4$ and the remaining scores are also equally likely. X is the drv 'the number of "ones" seen'.

2 In the distribution function of the drv X shown, calculate the value of d, which is a positive integer, given $E(X) = 2.5$.

x	1	2	d	4	$(d^2 - 4)$
$p(X = x)$	0.3	0.25	0.2	0.15	0.1

3 A die is biased so that $p(6) = \frac{1}{4}$, other scores being equally likely, and it is rolled three times. X is the drv 'the number of sixes seen in the three attempts'.

(a) Write down the probability distribution of X.

(b) What is the probability that three fives are seen in the three throws?

4 Two small unbiased cubes each have four faces painted red, one face painted white and one blue. The cubes are rolled together once and then a fair coin is tossed. If the coin comes down heads, then the number of red faces uppermost on the two cubes is recorded, but if the coin shows tails, the number of red faces is recorded as zero.

(a) Write out the probability distribution for the drv X, 'the number of red faces seen'.

(b) Calculate the value of $E(X)$.

5   The arrow on the spinner shown is
    equally likely to stop at any point
    round its circle. A biased coin
    {p(heads) = $\frac{1}{3}$} is thrown once and
    then the spinner is spun twice. If the
    coin shows heads, the number of times
    the arrow stops over the red sector is
    recorded, but if the coin shows tails,
    the number of times the spinner stops
    above the yellow sector is written down.
    Draw up probability distribution tables
    for the drv's X and Y where:

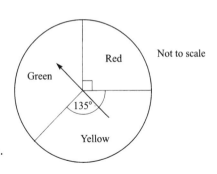

    X is 'the number of red sectors recorded when a head is thrown',
    Y is 'the number of yellow sectors recorded when a tail is thrown'.

6   Two cards are chosen at random from a full pack, with replacement.
    Write out the probability distributions for:

    (a) the number of aces seen,
    (b) the number of spades chosen,
    (c) the number of picture cards (king, queen, jack) seen.

7   (a) A group of four girls and six boys are the finalists in a scholarship
         competition. Three equal scholarships are to be awarded and, at this
         stage, each competitor is equally likely to win. Find the expected
         number of girls who win a scholarship and the variance of this
         distribution.

    (b) If, in the above competition, it is laid down that one scholarship is
         reserved for a girl, find now the expected number of girls who
         receive a scholarship.

8   Five pdf's are stated below for the drv X. Find E(X) and Var(X) for
    each example. K is a different constant in each case.

    (a) $x = 0, 1, 2, 3, 4$: $p(X = x) = (K + 1)x$
    (b) $x = 4, 5, 6, 7, 8$: $p(X = x) = K(x - 2)$
    (c) $x = 1, 2, 3, 4, 5$: $p(X = x) = \dfrac{K - 1}{x}$
    (d) $x = 2, 3, 4, 5, 6$: $p(X = x) = 2Kx.$
    (e) $x = -2, -1, 0, 1, 2$: $p(X = x) = Kx^2.$

9   A security lock on an office door has four buttons, one of which, when
    punched, opens the door. A forgetful employee punches one button after
    another, at random until he hits the right one. Once a button has been
    punched, it only becomes operative again after the door has been opened
    and shut. Find the expected number of buttons punched before the door
    is opened.

*10 The pdf of the drv X is defined as

$$p(X = x) = \begin{cases} k(x+1) & \text{for } x = 1, 2, 3, 4 \\ 2kx & \text{for } x = 5, 6, 7 \\ 0 & \text{otherwise} \end{cases}$$

where k is a constant. Calculate:

(a) k (b) $E(X)$ (c) $Var(X)$

11 The angles of the red, white and blue sectors of the spinner shown are $180°$, $60°$ and $120°$ respectively. The arrow is spun three times, and is equally likely to stop at any point of its circular motion.

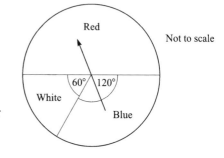

(a) Use a tree diagram to write out the pdf of the drv X, 'the number of times the arrow stops in the blue sector'.

(b) Calculate (i) $E(X)$, (ii) $Var(X)$.

(c) Write out the pdf and calculate the expectation and variance for the drv's Y and Z where:

Y is 'the number of times the arrow stops in the red sector',

X is 'the number of times the arrow stops in the white sector'.

*12 (a) A die is biased so that $p(3) = \frac{1}{2}$ and $p(6) = p(5) = p(4) = p(2) = p(1)$. The die is rolled twice. Write out the pdf for the drv X, 'the number of times the score is 6'. Calculate (i) $E(X)$, (ii) $Var(X)$.

(b) A fair die is rolled twice and the scores are added. Write out the pdf for the drv Y, 'the total of the scores'. Calculate (i) $E(Y)$, (ii) $Var(Y)$.

13 The pdf for the drv X is shown below.

$$p(X = x) = \begin{cases} kx^2 & x = 1, 2, 3 \\ 2kx & x = 4, 5, 6 \\ 0 & \text{otherwise} \end{cases}$$

where k is a constant. Calculate:

(a) $E(X)$ (b) $E(3X^2)$ (c) $E(2X + A)$,

where A is a constant

(d) $Var(2X - 1)$ (e) $E(3X^2 - 4X - 5)$

14 Six cards are taken from a full pack, two aces and all the kings. The cards are shuffled and placed face down on a table. A single card is chosen at random and not replaced. This is continued until the first ace is seen. The drv X is defined as 'the number of cards chosen up to and including the first ace'.

(a) Write out the pdf for X.
(b) Calculate (i) $E(X)$, (ii) the standard deviation of X.

15 The pdf of the drv X is shown below. You are given that $E(X) = 3.55$, and that X is positive. Calculate the value of x.

x	1	x	$x+1$	x^2	$x+3$
$p(X=x)$	0.05	0.15	0.25	0.3	0.25

Advanced

1 (a) The arrow shown will stop at random round the circular face, and it is spun once. Find the probability that it will stop over a red sector.
(b) If the arrow is spun twice, write down the probability distribution of the drv X, 'the number of times the spinner stops on a white sector'.
(c) For the double spin in (b), find (i) $E(X)$, (ii) $Var(X)$.
(d) If the arrow is spun three times, calculate $E(X)$ for this event, where X is 'the number of times the arrow stops over a white sector'.

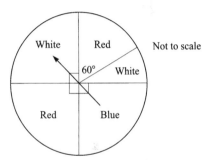

*2 ABC is a dartboard in the shape of an equilateral triangle. PQ is a straight line joining the mid-points of AB and AC respectively, and AR is a straight line with R the mid-point of BC. The numbers in the four areas are the scores made when the dart lands. A single dart is thrown at random and hits the board.

(a) Write out the pdf for the drv X, 'the score obtained by a single dart'.
(b) Calculate (i) $E(X)$, (ii) $Var(X)$.

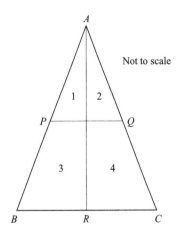

3 A student sits three examination papers for one of his A level subjects. Judging by past performance, the probabilities (independent of each other) of his passing are $\frac{4}{5}$ for the first paper, $\frac{2}{3}$ for the second and $\frac{1}{4}$ for the third.

(a) Find the probability that he passes two of the papers.
(b) Construct the pdf for the drv X, 'the number of papers passed'.
(c) Calculate (i) E(X), (ii) Var(X).

4 From a full pack of cards, all the diamonds and clubs are removed. From these 26 cards, four cards are removed, one after the other, at random and without replacement. D is the drv, 'the number of diamonds selected'.

(a) Write out the pdf for D.   (b) Show that E(D) = 2.

5 The pdf of the drv X is given as

$$p(X = x) = \begin{cases} k(x^2 + 2) & x = 0, 2, 4, 6 \\ kx & x = 8, 10, 12 \\ 0 & \text{otherwise} \end{cases}$$

(a) Calculate the value of k.
(b) Write out the pdf of X.
(c) Find (i) E(X), (ii) E($2X$), (iii) Var($2X + 5$).

6 The pdf of the drv X is shown below.

x	1	2	4	$2A$	$3A$	$5A$
$p(X = x)$	0.5	0.2	0.12	0.1	0.04	0.04

Calculate:

(a) the value of A, given E(X) = 2.94,
(b) Var(X),
(c) Var($5X + 3$).

7 A man goes regularly to a barber for a haircut. The time he has to wait for service depends on the number of customers already in the shop, and he makes a careful record of these numbers. From these figures, he calculates the probability of seeing various numbers of customers already in the shop when he gets there. The following table shows his results, where N, 'the number of customers in the shop when he arrives', is a drv.

n	0	1	2	3	4
$p(N = n)$	0.12	0.37	0.22	0.19	0.1

From these figures, calculate (a) E(N), (b) the standard deviation of N.

8 Aishaa is a notorious truant at school. Her class teacher keeps a record of the number of days each week that she is absent without good reason and, over a period of several weeks, is able to draw up a distribution table for the drv D, 'the number of days per week she plays truant'.

d	0	1	2	3	4
$p(D = d)$	0.12	0.16	0.43	0.24	0.05

(a) Calculate (i) $E(D)$, (ii) the standard deviation of D.
(b) During a term of eleven weeks, in how many weeks would you expect Aishaa to miss two days at school, to the nearest week? In a school year of 36 weeks, on how many days, to the nearest day, would you expect her to be playing truant?

9 Jane removed the labels from four cans of rhubarb and three cans of tuna fish, for use in a competition. She forgot to mark the cans which, without their labels, look identical. She now opens successive cans, chosen at random, looking for a can of tuna fish. If X represents the number of cans she has to open until she finds a can of tuna fish, then you are given that

$$P(X = r) = k(6 - r)(7 - r) \quad \text{for } r = 1, 2, 3, 4, 5.$$

(i) Copy and complete the table below. Show that $k = \frac{1}{70}$.

r	1	2	3	4	5
$P(X = r)$		20k		6k	

(ii) Draw a sketch to illustrate the probability distribution. State the modal value, and whether the distribution is positively or negatively skewed.
(iii) Find the mean and standard deviation of X.
(iv) A can of rhubarb costs 50 pence and a can of tuna fish costs £1.20. Find the expected total cost of the cans Jane has to open to find a can of tuna fish.
(v) Use a probability argument to show that the given formula is correct when $r = 3$.

[MEI]

Revision

1 Find which of the following examples defines a pdf for the drv X. Give numerical reasons for your conclusions.

Values of x	$p(X = x)$
(a) 1, 2, 3, 4	$\frac{1}{5}(x - 1)$
(b) 2, 4, 6	$\frac{1}{52}(x^2 - 1)$
(c) 5, 10, 15, 20	$\frac{1}{18}\left(\dfrac{x + 10}{5}\right)$
(d) 3, 5, 8, 9, 11	$\frac{1}{27}(x - 2)$
(e) $\frac{1}{2}, \frac{1}{4}, \frac{1}{8}, \frac{1}{16}$	$\dfrac{1}{30x}$

2 From two full packs of cards, all eight kings are removed. This small pack of 16 cards is shuffled, and then two cards are taken out, at random and without replacement.

(a) Write out the probability distribution for the drv X, 'the number of red kings seen'.
(b) If the selection were carried out with replacement, what would now be the probability distribution of X?

3 Two probability distributions, for the drv's X and Y, are given below.

x	0	1	2	3
$p(X = x)$	$\dfrac{1}{5}$	$\dfrac{2}{5}$	$\dfrac{1}{5}$	$\dfrac{1}{5}$

y	0	1	2	3
$p(Y = y)$	$\dfrac{A}{12}$	$\dfrac{A}{8}$	$\dfrac{A}{6}$	$\dfrac{A}{24}$

You are given that $E(Y^2) = 2E(X)$. Calculate the value of A.

*4 A pupil is entered for a piano grade examination. She finds one of the set pieces very awkward and often hits a wrong note. Her teacher does some calculations and finds a probability distribution for the drv M, 'the number of wrong notes hit' as follows:

m	0	1	2	3	4
$p(M = m)$	$\dfrac{1}{10}$	$\dfrac{4}{10}$	$\dfrac{2}{10}$	$\dfrac{2}{10}$	$\dfrac{1}{10}$

(a) Find, from these figures, (i) the number of wrong notes the pupil might be expected to hit in the examination, (ii) the variance of M.
(b) The pupil plays the piece 38 times in preparation for the examination. How many of these performances would you expect to be entirely free of mistakes (to the nearest whole number)?

5 Brian and Mia play a game where Brian spins a fair coin and Mia rolls a fair die. If the coin shows heads, Mia records the score on the die, but if it shows tails, 1 is subtracted from the score. Write out the probability distribution for the drv X, 'the score recorded'. Calculate

(a) $E(X)$,   (b) $Var(X)$,
(c) $p(\text{the recorded score} > 3)$,   (d) the total expected after 20 games.

6 A gardener has a passion for growing trees from seed. He plants seeds in groups of three and keeps a record of the number which germinate in each group. After a long period, he is able to write down the frequency distribution for the number of seeds which germinated from 120 plantings of three seeds.

No. germinating	0	1	2	3
Frequency	5	23	54	38

(a) Construct a probability distribution from these figures for the drv G, 'the number of seeds which germinate from a planting of three seeds'.

(b) Calculate (i) $E(G)$, (ii) the standard deviation of G.

7 A tetrahedral die has the scores 1, 2, 3, 4 on its faces. The score which counts when the die is rolled is the score on the face in contact with the table. The die is rolled 200 times, giving the following record.

Score	1	2	3	4
Frequency	36	44	48	72

(a) From these figures, write down the probability distribution for the drv S, 'the score when the die is rolled'.

(b) Calculate $\text{Var}(S)$.

8 A bag contains 12 discs, of which 9 are black and 3 are white. Discs are removed at random, and without replacement, until a black disc is seen. X is the drv 'the number of discs removed up to and including the seeing of a black disc'.

(a) Write down the pdf of X.

(b) Calculate $E(X)$ and $\text{Var}(X)$.

9 A high-jumper is allowed three attempts at each height of the bar. Near the end of the competition, the athlete becomes tired and the probability that she will clear the bar goes down with each attempt, as shown in the table. This is the probability distribution of the drv X, 'the number of attempts at which she fails to clear the bar when it is 2 cm below her personal best'.

x	0	1	2	3
$p(X = x)$	0.11	0.18	0.29	0.42

Calculate the expected number of failures she will make at this height.

10 (a) A poker die has on its six faces an ace, a king, a queen, a jack, a ten and a nine. Two fair poker dice are rolled together once. Write down the pdf of the drv N, 'the number of aces seen', and find $E(N)$.

(b) Two other dice each have four aces and two kings on their faces. They are both fair and are rolled together once. Find the pdf of the drv K, 'the number of kings seen', and find $E(K)$.

11 The probability distributions of three drv's X_1, X_2 and X_3 are as shown.

x_1	0	2	4	6
$p(X_1 = x_1)$	$\frac{1}{8}$	$\frac{2}{8}$	$\frac{2}{8}$	$\frac{3}{8}$

x_3	0	2	5	9
$p(X_3 = x_3)$	$\frac{1}{6}$	$\frac{1}{6}$	$\frac{2}{6}$	$\frac{2}{6}$

x_2	0	3	6	9
$p(X_2 = x_2)$	$\frac{1}{5}$	$\frac{1}{5}$	$\frac{1}{5}$	$\frac{2}{5}$

Calculate (a) $E(X_1)$, $E(X_2)$ and $E(X_3)$, (b) $\text{Var}(X_1)$, $\text{Var}(X_2)$ and $\text{Var}(X_3)$.

*12 This symmetrical trapezium is
a dartboard with the scores
as shown for the different areas
of the board. Two darts are
thrown at random and both
hit the board. Construct the pdf
of the drv S, 'the total score
made by the two darts'.

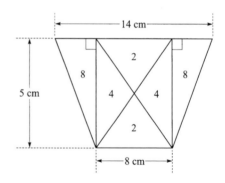

3.2   Binomial distribution

> Binomial distribution $E(X) = np$, $Var(X) = npq$
>
> $B(n, p) = (q + p)^n$, where $q = 1 - p$

Basic

1   For each part of this question, a different definition of the distribution of
the discrete random variable X is given. Calculate the required
probabilities.

    (a) X is $B(5, \frac{1}{4})$; find   (i) $p(X = 5)$,   (ii) $p(X = 0)$,   (iii) $p(X = 3)$,
         (iv) $E(X)$,   (v) $Var(X)$.

    (b) X is $B(8, \frac{1}{2})$; find   (i) $p(X = 8)$,   (ii) $p(X = 0)$,   (iii) $p(X = 5)$,
         (iv) $E(X)$,   (v) $Var(X)$.

    (c) X is $B(6, \frac{1}{3})$; find   (i) $p(X = 1)$,   (ii) $p(X \geq 2)$,   (iii) $p(X = 5)$,
         (iv) $E(X)$,   (v) the standard deviation of X.

    (d) X is $B(7, 0.3)$; find   (i) $p(X = 2)$,   (ii) $p(X = 4)$,   (iii) $p(X = 6 \text{ or } 7)$,
         (iv) $E(X)$,   (v) $Var(X)$.

    (e) X is $B(5, 0.2)$; find   (i) $p(X = 1)$,   (ii) $p(X > 1)$,   (iii) $p(X < 5)$,
         (iv) $E(X)$,   (v) the standard deviation of X.

    (f) X is $B(10, 0.6)$; find   (i) $p(2 \leq X < 4)$,   (ii) $p(X > 8)$,   (iii) $p(X = 6)$,
         (iv) $E(X)$,   (v) the standard deviation of X.

    (g) X is $B(7, 0.25)$; find   (i) $p(X = 2 \text{ or } 3)$,   (ii) $p(X < 1)$,
         (iii) $p(5 \leq X < 7)$,   (iv) $E(X)$,   (v) $Var(X)$.

    (h) X is $B(5, \frac{3}{4})$; find   (i) $p(X > 3)$,   (ii) $p(X \leq 2)$,   (iii) $p(X = 4 \text{ or } 5)$,
         (iv) $E(X)$,   (v) the standard deviation of X.

    (i) X is $B(6, \frac{1}{2})$; find   (i) $p(2 < X < 5)$,   (ii) $p(0 < X < 2)$,   (iii) $E(X)$,
         (iv) $Var(X)$.

    (j) X is $B(5, k)$; find   (i) $p(X = 5)$,   (ii) $p(X = 4)$,   (iii) $p(X = 1)$,
         (iv) $E(X)$,   (v) $Var(X)$.

    (k) X is $B(N, p)$; find   (i) $p(X = 1)$, (ii) $p(X = N - 1)$,   (iii) $E(X)$,
         (iv) $Var(X)$.

    (l) X is $B(2N, \frac{1}{2}p)$; find   (i) $p(X = 1)$,   (ii) $p(X = 2N - 1)$,   (iii) $E(X)$.

    (m) X is $B(10p, p)$; find   (i) $p(X = 1)$,   (ii) $p(X = 10p - 1)$,   (iii) $Var(X)$.

2 For $X = B(10, \frac{1}{4})$, calculate

(a) $p(X = 4)$ (b) $p(X \leq 1)$ (c) $p(5 \leq X < 8)$
(d) $E(X)$ (e) $Var(X)$.

3 Ballpoint pens are made in large quantities in a factory. They are tested in batches of 50 and it is known that 3% are faulty. From a test batch, calculate

(a) $p(X = 2)$ (b) the probability that none of the pens is faulty.

In 200 batches of 50, how many pens would you expect to be faulty?

4 In each of the following cases, write out the full probability distribution of the discrete random variable X. For the numerical answers, check that the sum of the probabilities is equal to 1.

(a) $X = B(5, 0.2)$

(b) $X = B(4, \frac{1}{4})$ [work in fractions here, not decimals]

(c) $X = B(4, 0.7)$ (d) $X = B(3, 1 - A)$ (e) $X = B(3, 2p)$

5 A cubical die is biased so that $p(6) = 0.6$. The die is rolled five times.

(a) Write out the probability distribution for the discrete random variable X, 'the number of sixes thrown on the die'. Give all the figures shown on your calculator display.

(b) Calculate (i) p(four sixes are thrown), (ii) p(only the first score is a six).

*6 A fair die is rolled six times. What is the probability that:

(a) no sixes are seen,

(b) exactly two sixes are shown,

(c) only the second and third rolls give a six?

7 A fair tetrahedral die, with its four faces marked 1, 2, 3 and 4, is rolled three times. Calculate the probability that:

(a) the total of the scores is 12,

(b) exactly two ones are seen,

(c) no fours appear.

8 The probability that a darts player will score 20 with any one dart is 0.6. He throws three darts at the board and none of them misses. Calculate the probability that:

(a) each of the three darts scores 20,

(b) only one dart scores 20,

(c) the player misses the 20 with one dart.

Show that, for this distribution, $5 \, Var(X) = 2E(X)$, where X is the discrete random variable 'the number of darts scoring 20' in a single 'throw' of three darts.

9 A factory produces bulbs for car headlights. The probability that any one bulb is defective is $\frac{1}{50}$ and they are packed in boxes of ten. From a single box of bulbs, find the probability that:

(a) none of the bulbs is defective,
(b) exactly two bulbs are defective,
(c) more than eight bulbs work properly.

10 Hockey balls are produced in large numbers in a factory and are despatched to a shop in boxes containing six balls. The probability of any one ball being defective is 0.15. A single box is checked. Find the probability that:

(a) none of the balls is defective,
(b) half the balls are defective,
(c) less than two balls are defective.

11 Boxes of matches are advertised as containing 48 matches. When a check is made on a large number of boxes, it is found that 2% contain fewer than 48 matches. The matches are sold in packets of ten boxes and a single packet is checked. Calculate the probability that:

(a) five boxes contain fewer than 48 matches,
(b) all the boxes contain 48 matches,
(c) 10% of the boxes have fewer than 48 matches.

Intermediate

1 In a pottery, plates are checked in batches of ten to see if they are of first or second quality. The probability of any one plate being a 'second' is $\frac{1}{8}$. Find, for a single batch of ten, the probability that:

(a) one plate is a 'second',
(b) there are between two and four plates, inclusive, which are 'seconds',
(c) 80% of the plates are of 'first' quality.

2 A die is biased so that $p(6) = \frac{1}{3}$. The die is rolled five times.

(a) Find the probability that:

   (i)   three sixes are thrown,
   (ii) the total of the scores is less than or equal to 29,
   (iii) a six is scored more often than any other number.

(b) Twenty people each roll this die five times. How many sixes would you expect to see?

3 A class of 20 students take a mathematics examination. The probability of any one student passing is $\frac{3}{5}$.

(a) Find the probability that:

   (i)   two students fail,

(ii)   75% of the students pass,

(iii)   equal numbers of the students pass and fail.

(b)   Ten classes, each containing 20 students, take this examination. How many would you expect to fail?

*4   At a certain holiday resort, records show that, in the first week of August, the probability of it raining on any one day is $\frac{1}{4}$. A family takes a holiday at this resort for this week. Find the probability that, in these seven days:

(a)   it rains on fewer than two days,

(b)   it is fine for five or more days,

(c)   only the last two days are fine,

(d)   four consecutive days are spoiled by rain.

5   A car manufacturer produces 300 cars of a particular type per day. Quality control ensures that the proportion of finished cars with faulty windscreen wipers is 1 in a 100. Ten cars from one day's production are delivered to a sales outlet.

(a)   Find the probability that in this batch:

(i)   exactly two cars have faulty windscreen wipers,

(ii)   more than eight cars have wipers which work properly.

(b)   A working week is five full days. If all the cars are sent out in batches of ten, calculate from the figures above how many cars would be expected to have faulty wipers in one week.

6   (a)   A family has five children. Assuming that boys and girls are equally likely to be born, find the probability that there are three boys and two girls in this family.

(b)   If the national proportion of boys to girls was $2:3$, what would then be the calculated probability of three boys and two girls in this family?

7   The discrete random variable X has a binomial distribution $B(n, 0.2)$. Find the value of n for which

(a) $p(X = 0) = 0.4096$,   (b) $p(X > n - 1) = 0.008$.

8   A discrete random variable X has a binomial distribution $B(8, 0.6)$.

(a)   Calculate (i) $p(X < 3)$, (ii) $p(X \geq 7)$.

(b)   If $p(3 \leq X < a) > 0.635$, find the minimum possible value of a.

9   A coin is biased so that $p(\text{heads}) = 0.6$. The coin is spun five times. Calculate the probabilities of the following results:

(a)   three heads are seen,

(b)   only the second throw is a head,

(c) more heads than tails are thrown,
(d) an even number of heads is recorded (zero is not taken as an even number),
(e) three consecutive tails are thrown.

10 England are to play Ruritania in five cricket test matches in the summer of the year 3000. The coin which is tossed to give one side the initial batting or fielding choice is biased so that p(tails) = 0.7. The England captain tosses the coin every time and the Ruritanian captain invariably calls heads.

(a) What is the probability that the Ruritanian captain wins the toss
(i) twice, (ii) five times, (iii) only for the first and last tests?
(b) What would these probabilities be if the coin were fair?

11 The discrete random variables X_1 and X_2 are distributed as $X_1 = B(n_1, p_1)$, $X_2 = B(n_2, p_2)$. You are given that

$$E(X_1) = \tfrac{3}{2}, \quad E(X_2) = 3, \quad \mathrm{Var}(X_1) = \tfrac{3}{4}, \quad \mathrm{Var}(X_2) = \tfrac{3}{4}$$

Calculate the values of n_1, n_2, p_1, p_2.

*12 If X_1, X_2 are discrete random variables distributed as $X_1 = B(n, p)$, $X_2 = B(n + 1, 2p)$, show that:

(a) $E(X_2) = 2\{E(X_1) + p\}$
(b) $\mathrm{Var}(X_2) = E(X_2)(1 - 2p)$
(c) $\mathrm{Var}(X_2) = 2\,\mathrm{Var}(X_1) + 2p(1 - np - 2p)$

13 A mini-lottery requires players to choose four numbers from one to ten inclusive. Show that 210 selections are possible. If 500 people make one independent selection each, find the probability of there being exactly two winners.

14 An unbiased coin is tossed five times. What is the probability that three heads are thrown? If five people each toss the coin five times, what is the probability that each throws three heads?

15 A factory produces car headlight bulbs and 2% of the bulbs are faulty. What is the largest sample of these bulbs that can be taken if the probability that none is faulty is greater than 0.6?

Advanced

1 The probability of scoring a head with a biased coin is p. The coin is tossed twice. Write out in full the binomial expansion which gives the probabilities of all the possible results of this experiment and explain the meaning of each term. The coin is now tossed three times. Write down, in terms of p, the probability that exactly two heads are obtained.

2 A man catches a bus to work on six days every week. We assume that the bus is always on time. The probability that the man is late for the bus on any one day is 0.3.

(a) Calculate the probability that, in any one working week of six days, the man (i) misses the bus three times, (ii) misses the bus on the second and third days only, (iii) misses the bus on exactly three consecutive days.

(b) Calculate the mean and variance of this distribution. On how many days in a working year of 50 weeks would you expect the man to miss the bus?

3 X_1 and X_2 are discrete random variables which are distributed as $X_1 = B(n, p)$, $X_2 = B(2n, 3p)$. Calculate $E(X_1)$, $E(X_2)$, $Var(X_1)$, $Var(X_2)$ and use these results to show that $Var(X_2) = 6\,Var(X_1) - 12pE(X_1)$.

4 A biased die is such that $p(6) = \frac{1}{4}$, but $p(X = 1) = p(X = 2) = \cdots = p(X = 5)$. When the die is rolled, a prize is won if the score is 1 or 6. Find the probabilities of winning 0, 1, 2 or 3 prizes for a single set of three rolls. If 50 people each roll the die three times, how many would you expect to win 0, 1, 2, or 3 prizes?

5 A coin is biased so that $p(\text{heads}) = p$. An experiment with the coin consists of tossing it four times and noting the number of heads shown. This experiment, repeated 200 times, gave the following results:

Number of heads	0	1	2	3	4
Frequency	1	10	40	86	63

Calculate the mean and variance of this distribution and hence find the value of p. Use this value of p to write out a binomial expansion from which theoretical values for the number of heads is calculated and compare these values with those found in the experiment.

*6 In a school with 800 pupils, one hundred are left-handed. One class in the school contains 20 pupils.

(a) Find the probability that in this class
   (i) five pupils will be left-handed,
   (ii) more than 17 will be right-handed.

(b) How many left-handed pupils would you expect to find in this class?
(c) If every class in the school contained 20 pupils, how many classes would you expect to contain three left-handed pupils?

7 The Army is targeting homeless people as possible recruits. Batches of 20 applicants are given tests and, on average, eight from each batch are accepted.

(a) In a given batch of 20, find the probability that:
- (i) ten applicants are accepted,
- (ii) more than two applicants are accepted,
- (iii) all the applicants are rejected.

(b) If 40 batches, each of 20 applicants, are tested, what is the most likely number of acceptances?

8 Wheel nuts for a particular type of car are made in large quantities in a factory. From a day's output, a batch of 40 nuts is checked for defects. If more than two nuts from the batch are defective, the day's output is rejected, and if fewer than two are defective, it is accepted. If just two nuts are defective, a second sample of 20 nuts is inspected. For the day's output to be accepted, all the nuts in this second sample have to be perfect. It is known that, over a long period, 1% of the nuts are defective. Calculate the probability that:

(a) the day's output is accepted at the first inspection,
(b) there are two defective nuts in the first batch, but none in the second,
(c) the day's output is rejected.

9 At a church fair, a game is played by taking a ball at random out of a bag. The bag contains three red, seven white and ten blue balls. A single 'go' consists of three removals, with replacement. A red ball scores 5 points, a white ball 2 points and a blue ball 1 point. Prizes are awarded as follows: £5 for a score of 15 points, £2 for 11 or 12 points, and £1 for 6 or 3 points.

(a) For a single 'go' of three removals, find the probabilities of winning (i) £5, (ii) £2, (iii) £1.
(b) A persistent competitor plays the game 10 times, trying to win the £5 prize. What is the probability that (i) he wins the £5 prize once, (ii) he wins the £2 prize twice, (iii) he wins the £1 prize three times? (Assume that no other prize is won in each case.)

10 Copies of an advertisement for a course in practical statistics are sent to mathematics teachers in a large city. For each teacher who receives a copy, the probability of subsequently attending the course if 0.09. Twenty teachers receive a copy of the advertisement. What is the probability that the number who subsequently attend the course will be (a) 2 or fewer, (b) exactly 4?

[AEB]

11 A study of the numbers of male and female children in families of a certain population is being carried out.

(a) A simple model is that each child in any family is equally likely to be male or female, and that the sex of each child is independent of the sex of any previous children in the family. Using this model calculate the probability that, in a randomly chosen family of 4 children,

(i) there will be 2 males and 2 females
(ii) there will be exactly one female given that there is at least one female.

(b) An alternative model is that the first child in any family is equally likely to be male or female, but that, for any subsequent children, the probability that they will be of the same sex as the *previous* child is $\frac{3}{5}$. Using this model calculate the probability that, in a randomly chosen family of 4 children,

(i) all four will be of the same sex,
(ii) no two consecutive children will be of the same sex.
(iii) there will be 2 males and 2 females.

[UCLES]

Revision

1 Find the greatest probability in each of the following binomial distributions.

(a) $X = B(100, 0.3)$ (b) $X = B(50, 0.5)$ (c) $X = B(30, 0.15)$
(d) $X = B(75, 0.804)$ (e) $X = B(100, 0.754)$

Give all the figures on your calculator display for each answer.

2 A factory manufactures felt-tip pens. The overall proportion of defective pens is found to be 0.01. A single batch of 200 pens is tested. Find the probability that in this batch:

(a) five pens are defective,
(b) more than 1% of the pens are defective.

What is the most likely number of defective pens that will be found in this batch?

3 Describe an experiment in which the terms of the binomial expansion of $X = B(5, 0.45)$ give the probabilities of the possible results. Explain clearly the meaning of each term of the expansion. Calculate the value of $E(X)$ and explain what it represents.

4 A pet hen lays a single egg every day. The probability that the egg is brown is b. Over a period of 5 days, the average number of brown eggs laid is three. What is the probability that in a randomly chosen 5-day period:

(a) none of the eggs is brown,   (b) four of the eggs are brown?

5 Refills for ballpoint pens are made in large quantities in a factory. The mean number of defective refills in the output has been reduced to 1 in

200. Each day, a batch of 50 refills is checked. Calculate the probability that on any one particular day:

(a) all the refills in the batch tested are perfect,
(b) 4% of the batch are defective,
(c) fewer than four refills in the batch are defective. Give all the figures shown on your calculator display for this answer.

6 The discrete random variable X has a binomial distribution such that $E(X) = 2$ and $Var(X) = 1.6$.

(a) If the distribution is written as $(q + p)^n$, where $q + p = 1$, find p, q, and n.
(b) For this distribution calculate (i) $p(X = 2)$, (ii) $p(X = 3$ or $7)$.

7 The proportion of people with blue eyes in a population is found to be 4%. A group of 250 people, chosen at random from this population, is checked. What is the most likely number of people with blue eyes that are found in the group?

8 Two discrete random variables X_1 and X_2 are distributed as $X_1 = B(n_1, p_1)$, $X_2 = B(n_2, p_2)$. You are given that (a) $8p_1 = 3p_2$, (b) $n_1 = 1 + n_2$, (c) $12E(X_1) = 5E(X_2)$, (d) $16Var(X_1) = 15Var(X_2)$. Find n_1, n_2, p_1, p_2.

9 A bag contains 20 coloured discs, of which eight are black and the rest are white. A single disc is taken out at random and then replaced. This is done six times. Write down the probability that:

(a) four white discs are drawn,
(b) more black than white discs are drawn,
(c) three consecutive black discs are drawn,
(d) only the final disc drawn is black.

10 A bag contains a large number of sweets, coloured either red or white. The proportion of red sweets to white sweets is $2:1$. Sweets are taken out of the bag at random, but are replaced (self-restraint!). Find the smallest number of sweets which must be removed if the probability of seeing at least one red sweet is greater than 0.99.

11 A class of 30 students sit a GCSE mathematics examination. Based on past performance, the probability of any one member of the class gaining an A grade is 0.3. Use the binomial distribution to find the probability that the following gain an A grade:

(a) 12 members,   (b) none of the set,
(c) fewer than four members of the set.

*12 A fair cubical die is rolled 50 times. Find the probability of:

    (a) throwing a five or a six on the first throw,
    (b) throwing a five or a six five times,
    (c) throwing a five or a six ten times.

3.3   Geometric distribution, Poisson distribution

> Geometric distribution $E(X) = \dfrac{1}{p}, \quad Var(X) = \dfrac{q}{p^2}$
>
> Poisson distribution:
>
> for Po(λ), $p(X = x) = \dfrac{e^{-\lambda}\lambda^x}{x!}, \quad E(X) = Var(X) = \lambda$

Basic

1 If $X = \text{Geo}(0.6)$, calculate:

    (a) $p(X = 2)$ (b) $p(X = 4)$ (c) $p(X > 3)$
    (d) $p(X > 5)$ (e) $p(X \le 3)$ (f) $p(X \le 6)$
    (g) $p[(X > 5)|(X > 2)]$ (h) $p[(X > 6)|(X > 4)]$ (i) $E(X)$
    (j) $Var(X)$ (k) the modal value of X

2 If $X = \text{Geo}(0.3)$, calculate the following quantities:

    (a) $p(X = 3)$ (b) $p(X = 5)$ (c) $p(X < 4)$
    (d) $p(X < 6)$ (e) $p(X \le 2)$ (f) $p(X \le 5)$
    (g) $p(X > 2)$ (h) $p(X > 4)$ (i) $E(X)$
    (j) the standard   (k) $p[(X > 7)|(X > 3)]$ (l) $p[(X > 5)|(X > 3)]$
         deviation of X

3 The arrow on the spinner is equally
likely to stop at any point round the
circle. A prize is given if it stops over
the shaded sector. Calculate the
probability that a prize is won

    (a) on the second throw,
    (b) before the fifth throw.

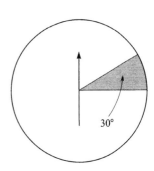

*4 A trout fisherman finds that the probability of catching a fish on any one cast of his line is 0.05. What is the probability that he catches a fish
  (a) on his third cast,   (b) after his 20th cast?

5 Claire and Tudor play a game of Snakes and Ladders, where a six must be thrown to start. Claire uses a fair die, but Tudor's die is biased so that $p(6) = \frac{1}{4}$. What is the probability that:
  (a) Claire starts on her third throw and Tudor on his fourth,
  (b) neither starts before the fourth throw?

6 For each of the following definitions of a Poisson distribution, calculate the required quantities.
  (a) Po(3):   (i) $p(X = 2)$ (ii) $p(X \leq 3)$ (iii) $p(X \geq 4)$ (iv) Var(X)
  (b) Po(2):   (i) $p(X = 3)$ (ii) $p(X \leq 2)$ (iii) $p(X = 4)$ (iv) Var(X)
  (c) Po(2.5): (i) $p(X = 5)$ (ii) $p(X < 3)$ (iii) $p(X = 2)$ (iv) the standard deviation of X
  (d) Po(4):   (i) $p(X = 4)$ (ii) $p(X > 2)$ (iii) $p(X \leq 3)$ (iv) the standard deviation of X
  (e) Po(2.1): (i) $p(X = 2)$ (ii) $p(X > 3)$ (iii) $p(X < 2)$ (iv) the standard deviation of X

7 Two variables, A and B, have Poisson distributions such that $A = $ Po(10), $B = $ Po(6). Calculate the following probabilities.
  (a) $p(A = 5$ or $6)$ (b) $p(A \leq 14)$
  (c) $p(9 \leq A \leq 12)$ (d) $p(B = 5)$
  (e) $p(B < 4)$ (f) $p(2 \leq B \leq 8)$
  (g) $p(A = 7$ and $B = 6)$ (h) $p(A < 10$ or $B < 6)$

8 If $X = $ Po(λ) and the standard deviation of X is 1.5, find
  (a) $p(X = 0)$,   (b) $p(X \leq 2)$.

9 If $X = $ Po(λ), and you are given that $E(X^2) = 5[E(X)]^2$, find
  (a) λ,   (b) $p(X = 0)$.

*10 Paper clips are made in a factory and the proportion of faulty clips in a day's production is known to be 4%. The number of faults follows a Poisson distribution. When a batch of 50 clips is tested, find the probability that in this batch:
  (a) less than 5 are faulty,   (b) none of the clips is faulty,
  (c) 6 or 7 or 8 are faulty.

11 In each question below, use the Poisson distribution as an approximation to the binomial to calculate the required probabilities.
  (a) $X = $ B(150, 0.05).
      Find (i) $p(X = 2)$, (ii) $p(X \leq 15)$, (iii) $p(8 \leq X \leq 12)$.

(b) $X = B(60, 0.08)$.
   Find (i) $p(X = 0)$, (ii) $p(X = 10)$, (iii) $p(2 \leq X \leq 4)$, (iv) $p(X = 15)$.
(c) $X = B(100, 0.09)$.
   Find (i) $p(X \leq 1)$, (ii) $p(X = 8)$, (iii) $p(X \leq 6)$.

12 In a gold mine, the number of accidents per shift follows a Poisson
distribution. For a given shift, the proportion of accidents to miners is
1%. During one particular shift, 400 miners work down the mine. Find
the probability that during this shift there are

(a) no accidents,   (b) fewer than two accidents,
(c) more than three accidents.

Intermediate

1 At a pottery, 10% of the output of a certain type of jug is thought to be
faulty. The finished pieces are stored side by side on long shelves, and
they have to be checked one by one. Find

(a) the expected number of jugs which have to be checked before a
   faulty one is found,
(b) the probability that the fifth jug tested is the first faulty one.

2 A tetrahedral die, with faces marked 1, 2, 3 and 4, is biased so that
$p(1) = \frac{1}{3}$, the other scores being equally likely. The die is thrown on to a
table and the score is taken as the number in contact with the table.
Calculate the probability that:

(a) the first 'one' is scored on the second throw,
(b) the first 'one' is scored on the tenth throw,
(c) the first 'one' is scored on either the third or fourth throws,
(d) no 'ones' are seen for the first five throws.

3 On the dartboard shown, all the areas
are equal. Assuming that all darts
(i) are thrown at random and hit
the board, (ii) are equally likely to
land in any square, find the probability
that:

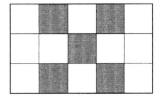

(a) any one dart strikes a shaded square,
(b) the fourth dart is the first to
   land in a shaded square,
(c) the first two darts land in an
   unshaded square.

Find the expected number of darts
thrown before one lands in a shaded square.

*4 Ye Ling and Sam play a game where each throws a single dart, in turn, at a conventional circular board. The first to throw a treble twenty wins the game. The probabilities that they hit the correct area are 0.3 (Ye Ling), and 0.5 (Sam). If Ye Ling always throws first, find the probability that:

(a) Sam wins on his second throw,
(b) Ye Ling wins on her second throw,
(c) Ye Ling wins the game.

5 The probability that it rains in Greyness on any one of the first 14 days in August is 0.2. If a holiday is booked at Greyness for these two weeks, find the probability that:

(a) the first rain falls on the fifth day,
(b) it rains only the first and last days of the holiday,
(c) there are four consecutive days of rain during the holiday, all the other days being fine.

6 Vicky and Jo play a game where they each toss the same coin in turn. For this coin, p(heads) = p, and the game is won when the first head is thrown.

(a) In the first game, Jo goes first but Vicky wins on her first throw. The probability that this happens is 0.25. Find the value of p.
(b) In the second game, Vicky goes first, but Jo wins at her first attempt. If the probability that this happens is $\frac{2}{9}$, find a value for p ($p > \frac{1}{2}$ in this case).
(c) Using $p = \frac{3}{5}$, calculate the probability that Vicky wins on her third attempt, if Jo goes first.

7 A drv X has the probability distribution

$$p(X = x) = \begin{cases} \dfrac{2}{3x}\left(\dfrac{1}{3}\right)^{x-1} & \text{for } x = 1, 2, 3, \ldots \\ 0 & \text{otherwise} \end{cases}$$

Find E(X).

8 A city centre car park has 120 spaces. It is found that the proportion of cars broken into during a particular hour of the day follows a Poisson distribution and is 3%. On one day, at this hour, the car park was $\frac{2}{3}$ full. Find the probability that, during this hour, the number of cars broken into was

(a) less than three,   (b) more than five.

9 When a car is taken for its annual MOT test, two of the items checked are brakes and headlights. On average, a particular garage finds the brakes and headlights to be faulty in 3% and 2.5% of cars respectively.

(a) This garage tested 30 cars on a Monday. Find the probability that:
(i) two of the cars had faulty brakes,
(ii) the headlights on fewer than two cars were defective.

(b) Find these probabilities, assuming
(i) the faults are distributed binomially,
(ii) the faults follow a Poisson distribution.

*10 The random variable X is such that $X = Po(\lambda)$. You are given that

$$p(X = 2) + \tfrac{1}{4}p(X = 3) = p(X = 4)$$

Calculate (a) λ, (b) $p(X = 4)$.

11 A newsagent buys ten copies of a magazine each week and, on average, sells eight of them. By:

(i) treating this as a binomial distribution,
(ii) using the Poisson formula,

calculate the probability that in a particular week he sells

(a) fewer than eight copies,   (b) more than eight copies.

12 John and Jehangir play a game of darts. John, who is not a good darts player, throws his darts at random, although they all hit the board. Jehangir is a more competent player, and he can throw a dart into the bulls eye on this board with probability 0.8, the bulls eye having an area 0.3 times the total board area. The game consists of alternate throws, and Jehangir goes first. The first player to land a dart in the bulls eye is the winner. Find the probability that:

(a) John wins on his first throw,
(b) Jehangir wins on his third throw,
(c) John wins the game.

13 A man goes away on an extended business trip and writes home to his wife. He is not a regular correspondent, the letters arriving at random at an average rate of one every three days. His wife, knowing that he will be away for 36 days, and being a statistician, calculates the probability that:

(a) she will receive two letters in the first twelve days,
(b) she will receive more than six letters in the first three weeks,
(c) she will receive no letters in the first nine days.

All letters may be assumed to be delivered by courier the day after posting, the first (it is hoped!) being written on the outward journey, and the 36 days starting the day after the man left home. What were the answers his wife calculated?

14 At the siege of Ladysmith in the Boer war, shells were fired into the town at random intervals during the day. One of the women in the town noted

the intervals between the arrivals of the shells. The average time between shellbursts was 6 minutes. Calculate the probability that:

(a) there were eight shellbursts in the first hour of a randomly chosen day,

(b) on another randomly chosen day, there were five bursts between noon and 12.30,

(c) in two randomly chosen hours of any one day, there were five bursts in each hour.

15 As part of complaints against a commuter rail company, a protester records the number of trains arriving late at a main railway station. This number is known to have a Poisson distribution with parameter λ, and it was calculated that the probability of no trains being late during any particular hour was 0.135 to three significant figures. Calculate:

(a) the value of λ (to three significant figures),

(b) the probability that fewer than three trains are late in a particular hour,

(c) the probability that, in three randomly selected hours, more than eight trains are late.

Advanced

*1 Packets of the popular breakfast cereal, Oatbix, contain pictures of well-known football players. There are five different pictures in the first series and, when all have been collected, a prize is offered by the manufacturer. The pictures are randomly distributed in the packets and every packet contains one picture.

(a) Find the probability that:

(i) the first three cards seen all carry the same picture,

(ii) each of the first four cards found carries a different picture.

(b) Two of the pictures are of the players Crasher and Dasher. What is the probability that the first two pictures are one of each of these players?

(c) Delphine has collected four of the five pictures. Find the smallest value of the integer N such that p(not more than N cards are needed to complete the set) > 0.8.

2 The rules of the popular board game, Slotopoly, require a 'one' to be thrown before a player can advance and see what each slot has to offer (one fair die only is used).

(a) For a single player, find the probability that:

(i) a 'one' is thrown at the second attempt,

(ii) four throws are made before a 'one' is seen (i.e. on the fourth attempt).

(b) Three players, A, B and C, play together, throwing the die in that order. Calculate the probability that:

(i) at the first attempt, only B throws a 'one',

(ii) all three players throw a 'one' at the second attempt.

(c) How many throws will player C need if the probability that he starts at the Nth throw is to be more than 0.1?

3 Two executives, Neela and Paul, make telephone calls. Neela makes an average of three calls in 30 minutes, while Paul's rate is five calls in the same time interval. Calculate the probability that:

(a) Neela makes five calls in a 30-minute period,

(b) Neela makes more than two calls in a 15-minute period,

(c) Paul makes fewer than two calls in a given interval of 20 minutes,

(d) between them, more than five calls are made in a period of 10 minutes.

4 At the Supergro garden centre, an outbreak of the voracious dark blue caterpillar is threatening a certain type of shrub. The number of caterpillars found on a single stem follows a Poisson distribution with $\lambda = 2$. If six or more caterpillars are found on a flower stem, the shrub cannot be sold.

(a) What is the probability that, on any one stem, six or more caterpillars are found?

(b) Twenty plants, chosen at random, are inspected. Find the probability that fewer than three plants have to be discarded as a result,

(i) using the binomial distribution formula,

(ii) using the Poisson approximation to the binomial.

5 A typist makes random errors which follow a Poisson distribution. The average count is of two errors per 400 words.

(a) A particular report consists of 2000 words. What is the probability that in this report there are:

(i) 5 errors,   (ii) 12 errors?

(b) The typist now types two reports of 1000 and 200 words respectively. What is the probability that:

(i) there are no errors in either report,

(ii) each report contains two errors,

(iii) there is a total of six errors in both reports?

6 (a) A firm making plastic rulers for schools inspects a large batch and finds that 2% of the rulers are faulty. Each day after this, a batch of 50 rulers is tested. Assuming the faults follow a Poisson distribution, find the probability that on any one day:

(i) three faulty rulers are found,

(ii) fewer than two rulers are faulty.

(b) The firm also makes protractors, and on a daily testing of 50 protractors, 3% are faulty. Assuming also a Poisson distribution for these faults, find the probability that on any one day:

(i)   one ruler and one protractor are faulty,
(ii) four rulers and no protractors are faulty,
(iii) there is a total of four faulty testings from both items.

7   A firm making metal washers for machinery tests them for (i) diameter and (ii) thickness. 2% are found to be faulty in diameter and 4% are faulty in thickness. The faults follow a Poisson distribution. Each day, twenty randomly chosen washers are tested. Find the probability that in a working week of five days:

(a) four washers have faulty diameters,
(b) five have thicknesses outside the maximum tolerance,
(c) four faults in all are found,
(d) more than one washer is defective in both diameter and thickness,
(e) two washers have faulty diameters and three have faulty thicknesses.

8   Frugal Bakeries claim that packs of 10 of their buns contain on average 75 raisins. A Poisson distribution is used to model the number of raisins in a randomly selected bun.

(a) Specify the value of the parameter.
(b) State any assumption required about the distribution of raisins in the production process for this model to be valid.
(c) Show that the probability that a randomly selected bun contains more than 8 raisins is 0.338.
(d) Find the probability that in a pack of 10 buns at least two buns contain more than 8 raisins.

[Edexel]

9   The number of births announced in the personal column of a local weekly newspaper may be modelled by a Poisson distribution with mean 2.4. Find the probability that, in a particular week,

(a) three or fewer births will be announced,
(b) exactly four births will be announced.

[AEB]

10 Travellers arrive at a railway station, to catch a train, either alone or in family groups. On an August Saturday afternoon, the number X, of travellers who arrive alone during a one minute interval may be modelled by a Poisson distribution with mean 7.5.

(a) Find the probability of six or fewer passengers arriving alone during a particular minute.

The number, Y, of family groups who arrive during a one minute interval may be modelled by a Poisson distribution with mean 2.0.

(b) Find the probability that three or more family groups arrive during a particular minute.

It is usual for one person to buy all the tickets for a family group. Thus the number of people, Z, wishing to buy tickets during a one minute interval may be modelled by $X + Y$.

(c) Find the probability that more than 18 people wish to buy tickets during a particular minute.

If four booking clerks are available, they can usually sell tickets to up to 18 people during a minute.

(d) State, giving a reason in *each* case, whether

  (i) more than four booking clerks should be available on an August Saturday afternoon,

  (ii) the Poisson distribution is likely to provide an adequate model for the total number of travellers (whether or not in family groups) arriving at the station in a one minute interval,

  (iii) the Poisson distribution is likely to provide an adequate model for the number of passengers, travelling alone, leaving the station, having got off a train, during a one minute interval.

(e) Give *one* reason why the model $Z = X + Y$, used in part (c), may not be exact.

[AEB]

Revision

*1 If $X = \text{Geo}(0.2)$, calculate:

(a) $p(X \leq 2)$ (b) $p(X \leq 4)$ (c) $p(X > 3)$
(d) $p(X > 1)$ (e) $p[(X > 3)|(X > 1)]$ (f) $p[(X > 4)|(X > 1)]$

2 The drv's X and Y are defined as $X = \text{Geo}(K)$ and $Y = \text{Po}(2)$. You are given that $E(X) \cdot p(Y = 2) = 4e^{-2}$. Calculate the value of K.

*3 Ahmed is a prolific scorer of goals in hockey. His record during a season is an average of two goals in every match of 70 minutes' playing time, the goals being scored randomly. In a group of three randomly chosen matches, calculate the probability that he will score:

(a) fewer than two goals,   (b) more than five goals.

Show that, correct to two significant figures, there is about a 10% chance that he will score eight goals in these three matches.

4 A firm making cheap drinking glasses produces an output in which 2% of the glasses are distorted and 3% are damaged. A delivery is being made up containing 144 glasses. What is the probability that:

(a) three or four glasses are distorted,
(b) more than two are damaged,
(c) two are faulty in both respects?

5 For the two variables $A = \text{Po}(5)$ and $B = \text{Po}(6)$, calculate the following probabilities:

(a) $\text{p}(A = 3)$ (b) $\text{p}(A \le 2)$ (c) $\text{p}(B = 7)$

(d) $\text{p}(B < 4)$ (e) $\text{p}(A + B = 10)$

6 A firm which manufactures ballpoint pens finds, after a thorough testing programme, that 2% of the total output of pens are defective. Fifty pens, chosen at random, are put on a flat surface for a further check. Calculate the probability that:

(a) there is one defective pen in the first five checked,

(b) the fourth pen is the first defective one found.

7 Three different binomial distributions are defined below. For the values of n given, calculate the required probabilities using

(i) the binomial formula,

(ii) the Poisson approximation to the binomial.

(a) $X = \text{B}(50, 0.02)$: $\text{p}(X_1 = 2)$ (b) $X_2 = \text{B}(120, 0.03)$: $\text{p}(X_2 = 3)$

(c) $X_3 = \text{B}(500, 0.01)$: $\text{p}(X_3 = 1)$

8 In Colorado Springs, in the USA, thunderstorms are generally experienced in the late afternoon in summer. A record of their frequency shows that they follow a Poisson distribution and that a storm occurs on average once every two days over a period of several weeks. For a period of ten consecutive days, chosen at random, calculate the probability that:

(a) no storms occur,

(b) there are seven storms,

(c) there are storms on more than half of the ten days.

9 Roger and Carolyn play a game where Roger rolls an 8-sided die (it is an octagonal prism) and Carolyn tosses a coin. The die is fair and has the numbers 2, 2, 4, 4, 4, 4, 8, 8 on its faces, but the coin is biased so that $\text{p}(\text{heads}) = \frac{1}{4}$. Carolyn first tosses the coin. If it shows heads, then the score on the die is doubled, but if it shows tails, then the score is recorded as shown.

(a) If the coin is tossed once, what is the probability that the recorded score will be

(i) 16,   (ii) 4?

(b) What is the probability that the first score of 16 will be on Carolyn's third toss of the coin?

10 A dragonfly rests on a twig on the look-out for passing flies. When it sees a fly, it chases it to catch, if possible, a meal. An observer, watching the dragonfly, notes that its flights follow a Poisson distribution with an average spacing of 3 minutes.

(a) The observer watches the dragonfly continuously for 30 minutes. Find the probability that, in this time, the dragonfly makes:

(i) nine attacking flights,   (ii) twelve attacking flights.

(b) On another day, the observer watches for three-quarters of an hour. What is the probability that in this time the number of attacking flights is

(i) 12,   (ii) 13, 14, or 15?

11 You are given that $X = Po(\lambda)$ and $E(X^2) = 12$. Find

(a) λ,   (b) $p(X = 3)$.

12 In both parts of this question $X = Po(\lambda)$.

(a) If $p(X = 2) = \frac{4}{3}p(X = 4)$, calculate (i) λ, (ii) $p(X = \lambda)$.

(b) If $p(X = 5) = \frac{7}{6}p(X = 7)$, calculate (i) λ, (ii) $p(X = 5)$.

4

Continuous distributions

4.1   Density functions, uniform distribution

$$E(X) = \int xf(x)\,dx, \text{Var}(X) = \int x^2 f(x)\,dx - \{E(X)\}^2$$

$$F'(X) = p(X \leq x) = \int_{-\infty}^{x} f(t)\,dt, \quad F(x) = f(x)$$

Uniform distribution: for $R(a, b)$,

$$E(X) = \tfrac{1}{2}(a + b), \quad \text{Var}(X) = \tfrac{1}{12}(b - a)^2$$

Basic

Abbreviations used: pdf probability density function, $f(x)$
                    cdf cumulative density function, $F(x)$
                    crv continuous random variable, X

1 Show that each of the definitions of the function $f(x)$ given below defines a pdf for the crv X. Sketch each function.

(a) $f(x) = \begin{cases} x & 0 < x < \sqrt{2} \\ 0 & \text{otherwise} \end{cases}$

(b) $f(x) = \begin{cases} 6x & 1 < x < \dfrac{2}{\sqrt{3}} \\ 0 & \text{otherwise} \end{cases}$

(c) $f(x) = \begin{cases} \dfrac{x}{4} & 0 < x < \sqrt{8} \\ 0 & \text{otherwise} \end{cases}$

(d) $f(x) = \begin{cases} \dfrac{32}{43}\left(1 - \dfrac{3}{x^2}\right) & 2 < x < 4 \\ 0 & \text{otherwise} \end{cases}$

(e) $f(x) = \begin{cases} \dfrac{1}{\pi}(1 + \cos x) & 0 < x < \pi \\ 0 & \text{otherwise} \end{cases}$

*2 Given below are several definitions of $f(x)$, the pdf of the crv X. Find in each case the value of the constant k. Given a rough sketch of each function.

(a) $f(x) = \begin{cases} 2kx & 1 < x < 2 \\ 0 & \text{otherwise} \end{cases}$

(b) $f(x) = \begin{cases} kx^2 - 1 & 2 < x < 3 \\ 0 & \text{otherwise} \end{cases}$

(c) $f(x) = \begin{cases} kx + \dfrac{1}{x^2} & 1 < x < 3 \\ 0 & \text{otherwise} \end{cases}$

(d) $f(x) = \begin{cases} x - k & 0 < x < 2 \\ 0 & \text{otherwise} \end{cases}$

(e) $f(x) = \begin{cases} kx^2 & 4 < x < 5 \\ 0 & \text{otherwise} \end{cases}$

*(f) $f(x) = \begin{cases} 4k(x^2 - 1) & 2 < x < 5 \\ 0 & \text{otherwise} \end{cases}$

(g) $f(x) = \begin{cases} k\left(\dfrac{1}{x^2} + \dfrac{1}{x^3}\right) & 1 < x < 3 \\ 0 & \text{otherwise} \end{cases}$

(h) $f(x) = \begin{cases} k^2 x & 0 < x < 3 \\ 0 & \text{otherwise} \end{cases}$

3 $f(x) = \begin{cases} 3kx^2 & 0 < x < 2 \\ 0 & \text{otherwise} \end{cases}$

is the pdf of the crv X.
(a) Find the value of k.
(b) Calculate $p(X \geq 1)$.
(c) Calculate $p(0.5 \leq X \leq 1)$.
(d) Sketch $y = f(x)$.

4 The pdf for the crv X is defined as

$$f(x) = \begin{cases} 4k & -1 < x < 2 \\ 0 & \text{otherwise} \end{cases}$$

(a) Find the value of k.   (b) Sketch $f(x)$.
(c) Calculate $p(-1 \le X \le 1)$.

5 Giving a reason in each case, determine which of the following definitions of $f(x)$ make it a pdf.

(a) $f(x) = \begin{cases} \frac{1}{4}x & 1 < x < 3 \\ 0 & \text{otherwise} \end{cases}$

(b) $f(x) = \begin{cases} \dfrac{x^2}{8} & -2 < x < 4 \\ 0 & \text{otherwise} \end{cases}$

(c) $f(x) = \begin{cases} 4x^3 - 1 & 1 < x < 3 \\ 0 & \text{otherwise} \end{cases}$

6 Find the value of the constant a if

$$f(x) = \begin{cases} 8x & 1 < x < a \\ 0 & \text{otherwise} \end{cases}$$

is to be a pdf.

7 For each pdf shown below, find (i) $E(X)$ and (ii) Var (X).

(a) $f(x) = \begin{cases} \frac{2}{3}x & 1 < x < 2 \\ 0 & \text{otherwise} \end{cases}$

(b) $f(x) = \begin{cases} \frac{2}{5}(1 + 3x) & 0 < x < 1 \\ 0 & \text{otherwise} \end{cases}$

(c) $f(x) = \begin{cases} x + \frac{1}{2} & 0 < x < 1 \\ 0 & \text{otherwise} \end{cases}$

(d) $f(x) = \begin{cases} \frac{2}{3}(2 - x) & 0 < x < 1 \\ 0 & \text{otherwise} \end{cases}$

*8 In each part of this question you are given a pdf for the crv X. Find the value of k and hence the form of the cdf for each definition.

(a) $f(x) = \begin{cases} 5kx^2 & 1 < x < 3 \\ 0 & \text{otherwise} \end{cases}$

(b) $f(x) = \begin{cases} \dfrac{1 + 2x}{k} & 1 < x < 2 \\ 0 & \text{otherwise} \end{cases}$

(c) $f(x) = \begin{cases} \dfrac{2k}{x^3} & 1 < x < 4 \\ 0 & \text{otherwise} \end{cases}$

(d) $f(x) = \begin{cases} k^2 - 2x & 1 < x < 3 \\ 0 & \text{otherwise} \end{cases}$

9 For each cdf defined below, find the form of the pdf.

(a) $F(x) = \begin{cases} 0 & x \le 1 \\ x - 1 & 1 \le x \le 2 \\ 1 & x \ge 2 \end{cases}$

(b) $F(x) = \begin{cases} 0 & x \le 0 \\ x - \dfrac{x^2}{4} & 0 \le x \le 2 \\ 1 & x \ge 2 \end{cases}$

(c) $F(x) = \begin{cases} 0 & x \le 0 \\ \frac{1}{15}(x^2 + 2x) & 0 \le x \le 3 \\ 1 & x \ge 3. \end{cases}$

10 The cdf of the crv X is as defined:

$$F(x) = \begin{cases} 0 & x \le 1 \\ \frac{1}{8}(x^2 - 1) & 1 \le x \le 3 \\ 1 & x \ge 3 \end{cases}$$

Find (a) $f(x)$, (b) $E(x)$, (c) $Var(X)$. Leave answers to (b) and (c) as fractions in their lowest terms.

*11 The function

$$f(x) = \begin{cases} \dfrac{2x}{9} & 4 < x < 5 \\ 0 & \text{otherwise} \end{cases}$$

is the pdf of the crv X.

Find *(a) the median, (b) the 60th percentile, (c) the lower quartile for this distribution.

12 Five definitions of a uniformly distributed crv X are given below. For each definition, calculate (i) the pdf of X,   (ii) $E(X)$,   (iii) $Var(X)$, (iv) $p(X > 5)$.

(a) $X = R(2, 10)$ (b) $X = R(4, 7)$ (c) $X = R(-2, 6)$
(d) $X = R(2A, 5B)$ (e) $X = R(a^2, b^2)$

13 The crv X has a uniform distribution with pdf $f(x)$ such that

$$f(x) = \begin{cases} \dfrac{1}{6} & 1 < x < k \\ 0 & \text{otherwise} \end{cases}$$

Find (a) k,   (b) the cdf of X.

14 Each pdf below represents a crv X which is uniformly distributed. For each definition find the following quantities and draw a sketch of $f(x)$:

(i) the value of k,   (ii) $E(X)$,   (iii) $Var(X)$,   (iv) $F(x)$.

(a) $f(x) = \begin{cases} \dfrac{1}{5} & k < x < 11 \\ 0 & \text{otherwise} \end{cases}$

(b) $f(x) = \begin{cases} \dfrac{1}{3} & k - 1 < x < 11 \\ 0 & \text{otherwise} \end{cases}$

(c) $f(x) = \begin{cases} \dfrac{1}{8} & k + 2 < x < 20 - 2k \\ 0 & \text{otherwise} \end{cases}$

(d) $f(x) = \begin{cases} \dfrac{1}{4} & k^2 < x < 13 \\ 0 & \text{otherwise} \end{cases}$

15 A uniform distribution is defined as $X = R(2, 7)$.

(a) Write down its pdf.   (b) Find $E(X)$ and $Var(X)$.
(c) Calculate $p(X < 4)$.   (d) Calculate $p(4 < X < 5)$.

*16 The parts of this question define uniformly distributed crv's.
Find $E(X)$, $\text{Var}(X)$ and the listed probabilities for each part.

(a) $X = R(3, 7)$.
    Find   (i) $p(X > 5)$,   (ii) $p(4 < X < 7)$.

(b) $X = R(10, 12)$.
    Find   (i) $p(X > 10)$,   (ii) $p(10.5 < X < 11.5)$.

(c) $X = R(-1, 3)$.
    Find   (i) $p(-1 < X < 1)$,   (ii) $p(-0.7 < X < -0.3)$.

*(d) $X = R(-10, -6)$.
    Find   (i) $p(-9 < X < -7)$,   (ii) $p(-8.5 < X < -7.5)$.

(e) $X = R(3A, 5B)$.
    Find   (i) $p(X > 1)$,   (ii) $p(3 < X < 5)$.

(f) $X = R\left(\dfrac{1}{P}, \dfrac{1}{Q}\right)$.

    Find   (i) $p(X < K^2)$,   (ii) $p(X > 10)$.

17 The diagram shows the pdf of a uniformly
distributed crv X. Two values, X_1 and X_2
are selected at random

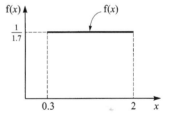

(a) Find the probability that
    (i) $X_1 < E(X)$,
    (ii) X_1 and X_2 are both greater than the
       standard deviation of X.

(b) Write down the pdf of X.
(c) Use symmetry to write down the median
    value of X.

18 The diagram shows the pdf of a crv X
which has a uniform distribution.

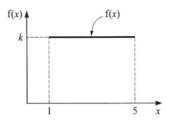

(a) Write down the value of k.
(b) Find $E(X)$ and $\text{Var}(X)$.

(c) Find   (i) $p\left(X > \dfrac{2}{\sqrt{3}}\right)$,

    (ii) $p[X - \text{Var}(X) > 2]$.

19 (a) The crv X is uniformly distributed as $X = R(A, 7)$. Given that
    $E(X) = 5$, calculate the value of A.

(b) The crv Y has a uniform distribution such that $Y = R(A, B)$, i.e.
    $B > A$. If $E(Y) = 1$ and $\text{Var}(Y) = 3$, calculate the values of A
    and B.

(c) Z is the crv $R(5, C)$, and you are given that $\text{Var}(Z) = \frac{3}{4}$. Calculate the
    value of C.

*20 The diagram shows the pdf of the uniformly distributed crv X. You are given that $E(X) = 1$.

Calculate   (a) A and B,

             (b) $p(X < 0)$.

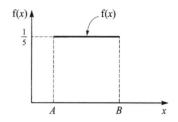

21 The figure displays the pdf of the uniformly distributed crv X. Given that $p(X < A^2) = \frac{1}{2}$, find (a) the positive value of A, (b) $Var(X)$.

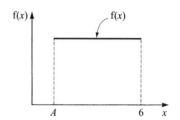

Intermediate

1   Show that each of the following definitions of the function $f(x)$ defines it as a pdf of the crv X.

(a) $f(x) = \begin{cases} \dfrac{x}{11} & 1 < x < 3 \\ \frac{1}{22}(x+3) & 3 < x < 5 \\ 0 & \text{otherwise} \end{cases}$

(b) $f(x) = \begin{cases} \dfrac{4x}{39} & 1 < x < 2 \\ \frac{2}{39}(x+2) & 2 < x < 5 \\ 0 & \text{otherwise} \end{cases}$

(c) $f(x) = \begin{cases} \frac{1}{22}(2x+3) & 1 < x < 2 \\ \frac{1}{22}(x+5) & 2 < x < 4 \\ 0 & \text{otherwise} \end{cases}$

(d) $f(x) = \begin{cases} \frac{3}{52}(1+x^2) & 1 < x < 2 \\ \frac{3}{52}(2x+1) & 2 < x < 4 \\ 0 & \text{otherwise} \end{cases}$

2   For each pdf defined below, find the value of k and the required probabilities.

(a) $f(x) = \begin{cases} 2kx & 1 < x < 2 \\ k(x+2) & 2 < x < 3 \\ 0 & \text{otherwise} \end{cases}$

Find (i) $p(X > 2)$, (ii) $p(\frac{3}{2} < X < 2)$.

(b) $f(x) = \begin{cases} kx^2 & 1 < x < 2 \\ k(3x - 2) & 2 < x < 3 \\ 0 & \text{otherwise} \end{cases}$

Find $p(2 < X < 3)$.

(c) $f(x) = \begin{cases} 3kx^2 & 1 < x < 3 \\ 27k & 3 < x < 4 \\ 0 & \text{otherwise} \end{cases}$

Find (i) $p(X < 2)$, (ii) $p(2 < X < 4)$.

*3   For each pdf defined below find (i) k, (ii) $E(X)$, (iii) $Var(X)$, (iv) $F(X)$.

*(a) $f(x) = \begin{cases} k(1 + x) & 1 < x < 2 \\ 0 & \text{otherwise} \end{cases}$

(b) $f(x) = \begin{cases} k(5 - x^2) & 1 < x < 2 \\ 0 & \text{otherwise} \end{cases}$

(c) $f(x) = \begin{cases} \dfrac{k}{x^2} & 1 < x < 3 \\ 0 & \text{otherwise} \end{cases}$

4   The crv X has its pdf defined below in different ways. Find the form of the cdf in each case.

(a) $f(x) = \begin{cases} 2x & 0 < x < 1 \\ 0 & \text{otherwise} \end{cases}$

(b) $f(x) = \begin{cases} \frac{1}{6}(1 + 2x) & 0 < x < 2 \\ 0 & \text{otherwise} \end{cases}$

(c) $f(x) = \begin{cases} \frac{3}{16}(4 - x^2) & 0 < x < 2 \\ 0 & \text{otherwise} \end{cases}$

(d) $f(x) = \begin{cases} \dfrac{3x^2}{7} & 1 < x < 2 \\ 0 & \text{otherwise} \end{cases}$

5 The pdf of the crv X is defined below:

$$f(x) = \begin{cases} \frac{1}{6}(3 + 2x) & 1 < x < 2 \\ 0 & \text{otherwise} \end{cases}$$

(a) Find (i) the median, m, (ii) the inter-quartile range, for this distribution.

(b) Find the form of $F(x)$ and the probability that X is greater than the expected mean of X.

6 The crv X has pdf $f(x)$, where

$$f(x) = \begin{cases} \frac{6}{37}(x + 1) & 1 < x < 2 \\ \frac{2}{37}(4x + 1) & 2 < x < 3 \\ 0 & \text{otherwise} \end{cases}$$

Find the complete description of $F(x)$.

7 For the pdf

$$f(x) = \begin{cases} k(4x - 1) & 2 < x < 4 \\ 0 & \text{otherwise} \end{cases}$$

calculate (a) the median, (b) the inter-quartile range (c) the 70th percentile value of X.

8 The time, T hours, taken to write a mathematics examination paper has pdf $f(t)$ given by

$$f(t) = \begin{cases} k(1 + t^2) & 0 < t < 1.5 \\ 0 & \text{otherwise} \end{cases}$$

(a) Calculate the value of k.

(b) Find the probability that a talented mathematician takes between 45 and 50 minutes to complete the paper.

(c) Show that, to two significant figures, the median time to complete the paper is 59 minutes.

*9 The pdf of the crv X is defined as $f(x)$, where

$$f(x) = \begin{cases} A + Bx & 0 < x < 4 \\ 0 & \text{otherwise} \end{cases}$$

and the median value of X is 3.

(a) Calculate the values of A and B.

(b) Check that your values of A and B make $f(x)$ a pdf.

10 The crv X has pdf

$$f(x) = \begin{cases} Ax - 2B & 0 < x < 10 \\ 0 & \text{otherwise} \end{cases}$$

The lower quartile value of X is 2. Calculate the values of A and B.

11 The crv X has pdf

$$f(x) = \begin{cases} 2Px - 3Qx^2 & 0 < x < 2 \\ 0 & \text{otherwise} \end{cases}$$

and its median value is 1.

(a) Find the values of P and Q.
(b) Calculate $E(X)$ and $Var(X)$.

12 The height, H m, jumped by the competitors at an international athletics meeting has pdf

$$f(h) = \begin{cases} 2kh & 1.8 < h < 2.2 \\ 0 & \text{otherwise} \end{cases}$$

(a) Show that $k = \frac{5}{8}$.
(b) Find the median height jumped, to the nearest centimetre.
(c) Show that, to the nearest centimetre, the expected mean height is the same as the median height.

13 The length, S km, of the line which can be written by the ink cartridge in a certain type of ballpoint pen has a pdf $f(s)$, where

$$f(s) = \begin{cases} 2s + K & 1 < s < 2 \\ 0 & \text{otherwise} \end{cases}$$

(a) Find the value of K.
(b) Find $E(S)$ and $Var(S)$.
(c) Calculate $p(S > 1.75\,\text{km})$.

14 The pdf of the crv X is defined as

$$f(x) = \begin{cases} k(9 - x^2) & 0 < x < 2 \\ 0 & \text{otherwise} \end{cases}$$

(a) Find (i) the value of k, (ii) $E(X)$, (iii) $Var(X)$.
(b) Show that the median value is approximately 0.9.

15 The crv X has a uniform distribution, $X = R(a, b)$.

(a) Given that $E(X) = \frac{1}{2}$ and $\frac{1}{2}(a + b) = b - a$, find (i) a and b, (ii) $Var(X)$.
(b) Draw a rough sketch to illustrate $f(x)$. Find $F(x)$.

16 The graph is the pdf of the uniformly distributed crv X.

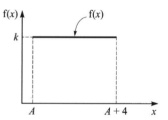

(a) Write down the value of k.
(b) Given that $p\{X < (A^2 + 1)\} = \frac{1}{4}$, $A \geq 0$, calculate (i) A, (ii) $E(X)$, (iii) $Var(X)$, (iv) $p(X < \{E(X) + Var(X)\})$.

*17 The crv X is uniformly distributed such that $X = R(A, A + 5)$.

(a) Given $E(X) = 3$, find the value of A.
(b) Find $p(X < \frac{4}{3})$.
(c) Show that $p(X > \{E(X) - Var(X)\}) = \frac{11}{12}$.

18 From the sketch of the pdf representing the crv X.

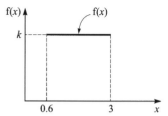

(a) write down the value of k,
(b) write out the pdf of X,
(c) find $F(x)$,
(d) calculate $F(2)$.

19 A circle has a radius of R cm, and this length is uniformly distributed over the interval 1–2.
If the circumference of the circle is C cm, find the pdf of C and calculate $E(C)$.

20 The crv X has a uniform distribution such that

$$f(x) = \begin{cases} \frac{1}{8} & 1 < x < 9 \\ 0 & \text{otherwise} \end{cases}$$

(a) Draw a rough sketch to illustrate $f(x)$.
(b) Find the following quantities.

  (i) $E(X)$ and $Var(X)$
  (ii) $p(X < 3)$
  (iii) $p\{E(X) < X < Var(X)\}$
  (iv) $p\{X > Var(X)\}$
  (v) A, if $p(A < X < A^2) = \frac{1}{4}$

21 X and Y are crv's defined as follows:

$$X = R(4, 10), \quad Y = 2X$$

Find $E(Y)$ and $Var(Y)$.

Advanced

1 The distance (D m) thrown by the javelin competitors at a Grand Prix event has a pdf f(D) such that

$$f(D) = \begin{cases} k(D+5) & 70 < D < 80 \\ 0 & \text{otherwise} \end{cases}$$

Find (a) the value of k, (b) E(D), (c) Var(D), (d) p($D > 75$).

2 A pdf is defined as follows for the crv X:

$$f(x) = \begin{cases} kx & 0 < x < 20 \\ k(40 - x) & 20 < x < 40 \\ 0 & \text{otherwise} \end{cases}$$

Calculate (a) the value of k, (b) E(X), (c) p($10 < X < 20$), (d) p($10 < X < 30$).

3 The crv X has a pdf f(x), such that

$$f(x) = \begin{cases} a + b(2x+1) & 0 < x < 2 \\ 0 & \text{otherwise} \end{cases}$$

and you are given that p($0 \le X \le 1$) = $\frac{2}{5}$.
Find (a) the values of a and b, (b) E(X), (c) Var(X).

*4 The pdf of the crv X is defined as

$$f(x) = \begin{cases} k(3x^2 - 2x) & 1 < x < 3 \\ 0 & \text{otherwise} \end{cases}$$

(a) Find (i) the value of k, *(ii) F(x), (iii) Var(X), (iv) p($X \ge 2$).
(b) Show that the median, m, is 2.5 to two significant figures.

5 A crv X has a pdf f(x), where

$$f(x) = \begin{cases} k(2x - 6x^2) & 1 < x < 3 \\ 0 & \text{otherwise} \end{cases}$$

Find (a) k, (b) E(X), (c) A, if p($A \le X \le 3$) = $\frac{81}{88}$, using a trial and error method.

6 The function f(x) as defined below is the pdf of the crv X.

$$f(x) = \begin{cases} k(4 - x^2) & 0 < x < 1 \\ 3kx & 1 < x < 2 \\ 0 & \text{otherwise} \end{cases}$$

(a) Calculate (i) the value of k, (ii) $E(X)$, (iii) $Var(X)$, (iv) $F(x)$.
(b) Show that m, the median value of X, lies between 1.4 and 1.5.
(c) Find the probability that, if two values of X are chosen at random, both will lie in the interval $(\frac{1}{2} < X < \frac{3}{4})$

7 The pdf, $f(x)$, for the crv X is given as

$$f(x) = \begin{cases} k(1 - 2\sin x) & 0 < x < \dfrac{\pi}{2} \\ 0 & \text{otherwise} \end{cases}$$

Find (a) the value of k, (b) $E(X)$, (c) $F(x)$, and check that $F\left(\dfrac{\pi}{2}\right) = 1$.

8 The continuous random variable X has probability density function f given by

$$f(x) = \begin{cases} k(2 - x) & \text{for } 0 \leq x \leq 2 \\ 0 & \text{otherwise,} \quad \text{where } k \text{ is a constant} \end{cases}$$

(i) Find the value of k (ii) Find the cumulative distribution function of X.
The continuous random variable Y is given by $Y = 1 - \frac{1}{2}X$, Show that $p(Y < y) = y^2$, where $0 \leq y \leq 1$.
Deduce the probability density function of Y and hence, or otherwise, show that $E(Y) = \frac{2}{3}$.
[UCLES]

9 A piece of wire is 80 cm long. It is bent to form a parallelogram $ABCD$, A being at the center of the wire. $AB = X$ cm, and $BAD = 30°$. The length of AB follows a uniform distribution of the form $R(p, q)$. The area of the parallelogram is Y cm^2.

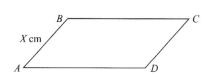

(a) Write down the values of p and q.
(b) Calculate (i) $E(X)$, (b) $E(X^2)$, (iii) $E(Y)$, (iv) $p(Y > 93.75)$.

10 The pdf of the uniform distributed crv X is

$$f(x) = \begin{cases} \frac{1}{10} & 1 < X < 11 \\ 0 & \text{otherwise} \end{cases}$$

and the uniformly distributed crv Y is such that $Y = 2X + 5$. Calculate $E(Y)$ and $Var(Y)$.

11 X is the uniformly distributed crv $R(2, 8)$. Y is also a uniformly distributed crv such that $Y = 3X - 2$.
Find (a) $E(Y)$, (b) $Var(Y)$, (c) $E(Y^2)$, (d) the pdf of Y, (e) $p(Y > 20)$.

12 X is the crv defined by the diagram,
You are given that $E(X) = 7$, and
$p(A + 3 < X < B - 1) = \frac{3}{5}$.
Find (a) A and B, (b) k,
(c) $E(Y)$ and Var(Y),
if $Y = 5X - 4$.

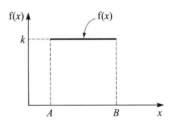

*13 X and Y are crv's. $Y = 3X$ and
X is defined by the diagram.
You are also given that
$E(X) = $ Var(X).
Calculate (a) A and B,
(b) Var(X), (c) $E(Y)$,
(d) Var(Y).

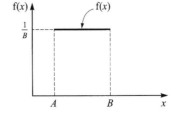

14 At an underground railway station, an average of two trains arrive every eight minutes.

(a) Find the probability density function which describes the length of time, T minutes, between the arrival of two consecutive trains.
(b) Find:
   (i) the mean time between trains,
   (ii) the cdf for T,
   (iii) the median time between trains.
(c) A student, late for an interview, runs onto the platform and finds that the previous train arrived two minutes ago. What is the probability that he will have to wait a further five minutes for a train?

15 (a) Two discrete random variables, X and Y, are such that

$$P(X = r) = \begin{cases} p_r & r = 1, 2, 3, 4, 5, \\ 0 & \text{otherwise} \end{cases}$$

and $P(Y = r) = \left(\frac{1}{r}\right)P(X \le r)$ for $r = 1, 2, 3, 4, 5$.

   (i) Show that $E(X) + E(Y) = 6$.
   (ii) Given that the distributions of both X and Y are symmetric, show that both X and Y have uniform distributions.

[AEB]

Revision

1 Sketch the following functions and, without performing any integration, determine which are pdf's for the crv X.

(a) $f(x) = \begin{cases} \dfrac{x+3}{3} & 0 < x < 2 \\ 0 & \text{otherwise} \end{cases}$

(b) $f(x) = \begin{cases} \frac{1}{6}(2x-1) & 1 < x < 3 \\ 0 & \text{otherwise} \end{cases}$

(c) $f(x) = \begin{cases} \frac{1}{18}(2x+1) & 1 < x < 4 \\ 0 & \text{otherwise} \end{cases}$

(d) $f(x) = \begin{cases} 6-2x & 1 < x < 2 \\ 0 & \text{otherwise} \end{cases}$

2 Show that the function $f(x)$ defined below is a pdf for the crv X.

$$f(x) = \begin{cases} \dfrac{1}{(x+1)\ln 4} & 0 < x < 3 \\ 0 & \text{otherwise} \end{cases}$$

3 For the function $f(x)$ defined below, show that it is a pdf only if $k = \dfrac{2}{\ln 5}$.

$$f(x) = \begin{cases} \dfrac{kx}{1+x^2} & 0 < x < 2 \\ 0 & \text{otherwise} \end{cases}$$

*4 Find $E(X)$ and $Var(X)$ for the definitions of $f(x)$ given below.

*(a) $f(x) = \begin{cases} \frac{1}{3}(4x+1) & 0 < x < 1 \\ 0 & \text{otherwise} \end{cases}$

(b) $f(x) = \begin{cases} \frac{2}{7}(x+2) & 1 < x < 4 \\ 0 & \text{otherwise} \end{cases}$

5 For each pdf defined below, find (i) the value of k, (ii) the median, (iii) $\text{Var}(X)$.

(a) $f(x) = \begin{cases} k(2x+1) & 0 < x < 2 \\ 0 & \text{otherwise} \end{cases}$

(b) $f(x) = \begin{cases} \frac{1}{2}k(x+1) & 0 < x < 1 \\ 0 & \text{otherwise} \end{cases}$

6 The crv X has its pdf defined in different ways below. Find the complete form of $F(x)$ for each definition.

(a) $f(x) = \begin{cases} \dfrac{3}{2x^2} & 1 < x < 3 \\ 0 & \text{otherwise} \end{cases}$

(c) $f(x) = \begin{cases} \frac{12}{29}\left(3 - \dfrac{x^2}{4}\right) & 1 < x < 2 \\ 0 & \text{otherwise} \end{cases}$

(c) $f(x) = \begin{cases} \frac{1}{9}(4 - x^2) & -1 < x < 2 \\ 0 & \text{otherwise} \end{cases}$

7 The crv X has pdf

$$f(x) = \begin{cases} x - k & 2 < x < 3 \\ 0 & \text{otherwise} \end{cases}$$

Find (a) the value of k, (b) the median value of X, (c) the 40th percentile, (d) $p(X < 2.5)$, (e) $p(2.2 < X < 2.4)$, (f) the lower quartile of this distribution.

8 The time, T hours, taken by several students at university to write a weekly essay is a crv with pdf

$$f(t) = \begin{cases} k(1 + t) & 0 < x < 2.5 \\ 0 & \text{otherwise} \end{cases}$$

Find (a) the value of k, (b) the probability that the time taken to write the essay by a randomly chosen student is less than 2 hours. (c) the expected mean time, to the nearest minute, taken for the essay.

9 For the cdf

$$F(x) = \begin{cases} 0 & x \le 2 \\ \frac{1}{5}(10x - x^2 - 16) & 2 \le x \le 3 \\ 1 & x \ge 3 \end{cases}$$

find (a) $F(2.4)$, (b) $p(X \ge 2.8)$, (c) $f(x)$, (d) $E(X)$.

10 The HOP brewery has a weekly delivery of X tonnes of hops, where X is a crv whose pdf is shown below.

$$f(x) = \begin{cases} k(4 - 5x) & 0 < x < 1 \\ 0 & \text{otherwise} \end{cases}$$

(a) Find the expected mass, in kilograms, of hops delivered each week.
(b) Find the cdf of this distribution.

11 The pdf of the crv X is defined as

$$f(x) = \begin{cases} 2kx(x^2 + 1) & 0 < x < 2 \\ 0 & \text{otherwise} \end{cases}$$

Find (a) k, (b) the median value of X, (c) Var(X).

12 For each part of this question, find $f(x)$, E(X), Var(X), and the probabilities which are requested for the given uniform distributions.

(a) R(15, 20): (i) p(16 < X < 19)   (ii) p(17.1 < X < 18.6)
(b) R(−4, 2): (i) p(−4 < X < −2)   (ii) p(−3 < X < 2)
    (iii) p{X < $\frac{1}{3}$ Var(X)}
(c) (i) p{X < E(X)}
    (ii) p(X > B)
    (iii) p(X > B + 12k)
(d) R(A − 8, A + 12):
    (i) p(X < 2A)
    (ii) p(X > 6)

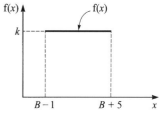

13 The sketch shows the pdf of a uniformly distributed crv X.

(a) Find k in terms of a and b.
(b) If E(X) = 4.5, and Var(X) = $\frac{3}{4}$, find a, b, and k.
(c) If p($X < C$) = $\frac{5}{6}$, find the value of C.

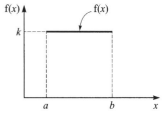

14 The crv X has a uniform distribution which is defined in the diagram.

(a) Find a relation between A and B.
(b) If p($X < 6.5$) = $\frac{5}{8}$, find the values of A and B.
(c) Shows that 9 Var(X) = 2E(X).

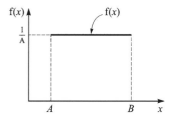

15 The crv X has a uniform distribution $X = R(4, 10)$. The crv Y is also uniformly distributed as $R(P, Q)$. You are given that $E(Y) = \frac{1}{2}E(X)$, and $Var(Y) = Var(X) - \frac{11}{12}$. Calculate the values of P and Q.

16 The crv X has a uniform distribution $X = R(10, 22)$. The crv Y is distributed as $Y = X + 2$. Calculate $E(Y)$ and $Var(Y)$.

17 A uniform distribution is described as $X = R(-3, 7)$.

(a) Given a rough sketch of this distribution.
(b) Calculate $E(X)$ and $Var(X)$.
(c) Find the following probabilities:

    (i) $p(X < 0)$
    (ii) $p(-1 < X < 4)$
    (iii) $p\{X > 3E(X)\}$
    (iv) $p(4.7 < X < 6.2)$

(d) If $p(X > A) = 0.25$, find the value of A.

18 The crv X has a uniform distribution such that $X = R(A, B)$. You are given that $E(X) = 4$, $Var(X) = \frac{49}{3}$, $B - A > 0$, $p(X > C) = \frac{1}{4}$, and $p(D < X < E) = \frac{1}{2}$, where $E + 5 = B$.
Find the values of A, B, C, D and E and given a rough sketch of the pdf of X.

4.2   Normal distribution

Basic

1 If $Z = N(0, 1)$, find:

(a) $p(Z < 1)$ (b) $p(Z < 0.6)$ (c) $p(Z < 2)$
(d) $p(Z < 0.25)$ (e) $p(Z < 0.94)$ (f) $p(Z < 1.37)$
(g) $p(Z < 0.576)$ (h) $p(Z < 1.028)$ (i) $p(Z < 2.345)$
(j) $p(Z < 2.758)$

2 If $Z = N(0, 1)$, find:

(a) $p(Z > 0.7)$ (b) $p(Z > 0.9)$ (c) $p(Z > 1.2)$
(d) $p(Z > 2.05)$ (e) $p(Z > 0.873)$ (f) $p(Z > 0.286)$
(g) $p(Z > 2.375)$ (h) $p(Z > 1.994)$ (i) $p(Z > 2.473)$
(j) $p(Z > 0.065)$

3 If $Z = N(0, 1)$, find:

(a) $p(Z > -0.254)$ (b) $p(Z > -1.045)$ (c) $p(Z > -0.761)$
(d) $p(Z > -0.496)$ (e) $p(Z > -2.462)$ (f) $p(Z > -1.507)$
(g) $p(Z > -2.085)$

4 If $Z = N(0, 1)$, find:

(a) $p(Z < -0.3)$ (b) $p(Z < -0.8)$ (c) $p(Z < -1.4)$
(d) $p(Z < -1.65)$ (e) $p(Z < -1.384)$ (f) $p(Z < -2.76)$
(g) $p(Z < -1.049)$

5 If $Z = N(0, 1)$, find:

(a) $p(-0.4 < Z < 0.7)$ (b) $p(-0.8 < Z < 1.1)$
(c) $p(-1.5 < Z < 0.38)$ (d) $p(-0.643 < Z < 1.025)$
(e) $p(-0.819 < Z < 2.462)$ (f) $p(-2.32 < Z < 0.276)$
(g) $p(-1.07 < Z < 1.971)$

6 If $Z = N(0, 1)$, find:

(a) $p(0.8 < Z < 1.3)$ (b) $p(0.27 < Z < 0.95)$
(c) $p(0.49 < Z < 0.75)$ (d) $p(0.556 < Z < 1.354)$
(e) $p(0.348 < Z < 1.027)$ (f) $p(1.24 < Z < 2.44)$
(g) $p(1.17 < Z < 1.86)$

*7 If $Z = N(0, 1)$, find:

*(a) $p(-0.8 < Z < -0.4)$ (b) $p(-0.6 < Z < -0.2)$
(c) $p(-0.73 < Z < -0.26)$ (d) $p(-1.5 < Z < -0.81)$
(e) $p(-1.23 < Z < -0.614)$ (f) $p(-0.769 < Z < -0.458)$
(g) $p(-2.52 < Z < -1.042)$

*8 If $Z = N(0, 1)$, find:

(a) $p(|Z| < 0.5)$ (b) $p(|Z| < 0.2)$ (c) $p(|Z| < 1.07)$
(d) $p(|Z| < 1.58)$ (e) $p(|Z| < 2.53)$ *(f) $p(|Z| > 0.35)$
(g) $p(|Z| > 0.82)$ (h) $p(|Z| > 1.13)$ (i) $p(|Z| > 1.64)$
(j) $p(|Z| > 2.01)$

9 For $Z = N(0, 1)$, find each of the following probabilities. Draw a rough sketch to show the area you are calculating in each case.

(a) $p(-0.7 < Z < 1.4)$ (b) $p(Z > 0.86)$
(c) $p(Z < -1.61)$ (d) $p(|Z| < 0.762)$
(e) $p(Z < 0.966)$ (f) $p(0.35 < Z < 0.645)$
(g) $p(|Z| < 0.906)$ (h) $p(-0.75 < Z < -0.33)$
(i) $p(|Z| > 1.14)$ (j) $p(Z > -0.585)$

10 For $Z = N(100, 100)$, calculate:

(a) $p(Z < 95)$ (b) $p(Z < 122)$ (c) $p(Z > 112)$
(d) $p(Z > 92)$ (e) $p(88 < Z < 104)$

11 For $Z = N(100, 100)$, calculate:

(a) $p(94 < Z < 98)$ (b) $p(|Z - 100| < 25)$
(c) $p(|Z - 100| < 16)$ (d) $p(|Z - 100| > 12)$
(e) $p(Z < 85)$

12 For $Z = N(50, 25)$, find k:

(a) $p(X < k) = 0.8962$ (b) $p(X < k) = 0.7454$
(c) $p(X < k) = 0.9803$ (d) $p(X < k) = 0.6145$
(e) $p(X < k) = 0.9180$

Intermediate

1 In each of the following, calculate the unknown parameter.

(a) $X = N(10, \sigma^2)$: $p(X < 12) = 0.8413$
(b) $X = N(8, \sigma^2)$: $p(X < 9.5) = 0.6915$
(c) $X = N(15, \sigma^2)$: $p(X < 19.6) = 0.9772$
(d) $X = N(0.6, \sigma^2)$: $p(X < 0.825) = 0.9332$

2 Calculate the value of the unknown parameter in each of the following.

(a) $X = N(-4, \sigma^2)$: $p(X < -5.47) = 0.0808$ (give this answer to 4 d.p.)
(b) $X = N(35, \sigma^2)$: $p(X > 39.5) = 0.0359$
(c) $X = N(0.05, \sigma^2)$: $p(X > 0.0577) = 0.0139$
(d) $X = N(100, \sigma^2)$: $p(X > 103) = 0.3538$

3 In each of the following, calculate the value of μ.

(a) $X = N(\mu, 2.25)$: $p(X < 8.75) = 0.9821$
(b) $X = N(\mu, 121)$: $p(X < 163.2) = 0.8849$
(c) $X = N(\mu, 23.04)$: $p(X < 83.4) = 0.9599$
(d) $X = N(\mu, 0.0064)$: $p(X < 0.374) = 0.9394$

4 Calculate the value of μ in each of the following.

(a) $X = N(\mu, 9.61)$: $p(X < 36.3) = 0.0139$ (give this answer to 4 s.f.)
(b) $X = N(\mu, 272.25)$: $p(X < 267.2) = 0.1151$
(c) $X = N(\mu, 1.96)$: $p(X < 15.08) = 0.9861$
(d) $X = N(\mu, 0.0441)$: $p(X > -2.736) = 0.9452$

5 Fifteen athletes compete in the triple-jump event at a Grand Prix meeting. The distances (D m) that they jump are normally distributed according to $D = N(17.1, 0.04)$.
In the first half of the competition, the athletes have to jump a distance of 17.15 m to qualify for the final section. How many athletes would you expect, the nearest whole number, to achieve this?

6 A crv X has a normal distribution $X = (5, \sigma^2)$. You are given that $p(X < 4) = 0.1056$. Show that the value of σ^2 is 0.64. Using this value, calculate:

(a) $p(X > 6.5)$ (b) $p(|X - 5| < 0.5)$

7 Two A level mathematics classes, totalling 38 students, are set a test. The marks are found to be normally distributed and the proportions of students scoring more than 70% and less than 40% are 15% and 10%

respectively. Calculate the mean and standard deviation of this distribution.

8 Safety matches are cut by a machine to a mean length of 4.2 cm. The lengths have a normal distribution with standard deviation 0.15 cm and the matches must be between 4.04 cm and 4.38 cm in length to be accepted. What percentage of the matches are rejected?

*9 A child's toy football has a diameter (D cm) which is normally distributed as $D = N(15, \sigma^2)$. The limits of diameter allowed are from 14.2 cm to 15.8 cm. If 3.8% of the balls are rejected, calculate a value for the variance of the diameter.

10 An engineering firm makes cast iron rods for a printing machine. The length of a rod, L cm, is normally distributed as $L = N(31, 0.0004)$. If 2% of the rods are rejected as being too short and 3% as being too long, what is the interval of acceptable lengths, to the nearest 0.001 cm, of these rods?

11 The masses of a group of men are distributed normally with mean 76 kg and variance 16 kg^2. Find the probability that a man, chosen at random from the group, will have a mass which is:

(a) less than 79 kg, (b) more than 72 kg,
(c) more than 82 kg, (d) between 72 kg and 78 kg,
(e) between 69 kg and 73 kg.

12 The speeds of cars, in km/hour, along a straight stretch of motorway are measured by traffic police and are found to be normally distributed as $V = N(115, 36)$.

(a) Find the probability that the speed of a single car, chosen at random, is (i) more than 130 km/h, (ii) between 116 km/h and 125 km/h, (iii) below the national speed limit of 112 km/h.
(b) Four randomly chosen cars are checked. What is the probability that only the first to be checked is exceeding the speed limit?

13 A crv X has a normal distribution defined as $X = N(\mu, 25)$. You are given that p$(X < 15) = 0.1587$. Find:

(a) the value of μ (b) p$(|X - \mu| < 8)$.

14 The crv X is such that $X = N(\mu, 16)$, and you are also given that p$(X < 25.8) = 0.8696$. Find the value of μ.

15 A crv X is normally distributed as $X = N(\mu, \sigma^2)$. Calculate the values of μ and σ^2, given that p$(X > 14.2) = 0.1357$, and p$(X < 9.4) = 0.0968$.

Advanced

*1   A professional long-jumper, when in training, lays out a target jump-spread of 6.7 m to 7.1 m. Over a period of 5 days' training, he makes 120 jumps (assuming sufficient rest between, and equal efforts at, each jump), 15 of which are longer than 7.1 m and 10 of which are less than 6.7 m. If we assume that the lengths are normally distributed with mean μ and variance σ^2, calculate μ and σ^2 to three significant figures. Of the 120 jumps, how many would you expect to be longer than 7.2 m?

2   Packets of ground coffee for use in a filter machine have a nominal mass of 250 g, and their standard deviation is 2.1 g. The masses of the packets can be assumed to be normally distributed. Quality control results in 4% of the packets being rejected as having too little coffee. What is the minimum acceptable mass of coffee in a packet?

3   The volume of beer in a bottle is stated to be 500 ml. This volume is normally distributed with mean 500 ml and variance $3.5 \, \text{ml}^2$.

   (a) Calculate the probability that the volume of beer in a randomly chosen bottle is more than 502 ml.
   (b) If the percentage of bottles containing less than V ml of beer is found to be 5%, calculate a value for V, to the nearest ml.

4   A bottle of chutney on a supermarket shelf has a declared mass of 335 g, and it is known that this mass (M g) is normally distributed with variance $81 \, \text{g}^2$. What is the inter-quartile range of M?

5   A supermarket buys small turkeys from a wholesale farmer and the masses (M kg) of the birds are normally distributed as $M = N(3, 0.0009)$. If 2% of the birds are rejected as being too light and 1.5% for being too heavy, find the acceptable mass limits, to the nearest gram, for each bird.

6   A factory manufactures glass tumblers whose masses (M g) are normally distributed according to $M = N(340, 8)$. Find the central symmetric interval within which 50% of the masses of the tumblers lie. If this range were from 337.5 g to 342.5 g, what would now be the value of the variance?

7   A salesperson in the furnishing department of a large store cuts many lengths of material each day. A check on the lengths (L m) sold as 1 m long shows that they follow a normal distribution with $L = N(1.01, \ 2.5 \times 10^{-5})$.

   (a) Find the probability that a randomly observed cutting gives a length of less than 1 m.

(b) If, during a single working day, 115 pieces of material are cut and sold as 1 m lengths, how many of them would you expect to be longer than 1.015 m?

(c) If the average cost of the materials sold in the department is £16.50 per metre, what would be an estimate of the cost, to the nearest penny, of the material supplied free to the customers in a week when 350 one-metre lengths of material are sold?

8   A work study report in a television component factory states that the times (T minutes) taken to complete the soldering of a component are normally distributed according to $T = N(12, 3.24)$. A randomly selected employee completes four of these components in a morning session of work.

(a) What is the probability that each piece will be made in less than 10 minutes?

(b) Find the following probabilities:

    (i) two of the four components are completed in 11 minutes each and the other two in 12.5 minutes each,

    (ii) none of the four components takes more than 11.6 minutes to complete.

9   The average proportion, p, of insects killed by the administration of x units of an insecticide is given by

$$p = P\left(Z \le \frac{x - \mu}{\sigma}\right)$$

where μ and σ are constants and Z is the normally distributed random variable with zero mean and unit variance.

When $x = 10$, $p = 0.4$ and when $x = 15$, $p = 0.9$.

    (i) Estimate, correct to three decimal places, the values of μ and σ.

    (ii) If a dose of 12.5 units is administered to 1000 insects, how many can be expected to survive

    (iii) What dose will be lethal to 99% of the insect population, on average?

[AEB]

Revision

1   If $X = N(100, 36)$, find the value of k in each of the following.

(a) $p(X > k) = 0.75$ (b) $p(X > k) = 0.38$ (c) $p(X > k) = 0.08$
(d) $p(X > k) = 0.12$ (e) $p(X > k) = 0.6$

2   If $X = N(50, 64)$, find k when:

(a) $p(X < k) = 0.02$ (b) $p(X < k) = 0.85$ (c) $p(X < k) = 0.45$
(d) $p(X < k) = 0.6$ (e) $p(X < k) = 0.9$

3 A box of household candles carries a claim that each candle will burn for
up to 5 hours. Investigation shows that the burning times (T hours) are
normally distributed as $T = N(4.7, 0.16)$. A box contains six candles.

(a) Find the probability that:

(i) a single, randomly selected candle will burn for less than
four hours.

(ii) each candle in a randomly chosen box will burn for more than
five hours.

(b) If it were found that an error had been made in calculating the mean
of the distribution, the new value being 4.8 hours, what would now
be the answer to (ii)?

4 X is a crv which is normally distributed as $X = N(\mu, \sigma^2)$, and
$p(X > 55) = 0.25$, $p(X < 37) = 0.2$

(a) Calculate (i) $p(X < 50)$, (ii) $p(X < 28)$, (iii) $p(40 < X < 50)$,
(iv) $p(|X - 47| < 5)$.

(b) If $p(|X - 47| < k) = 0.16$, calculate the value of k.

*5 The crv Y has a normal distribution such that $Y = N(4.5, \sigma^2)$, and
$p(Y > 5.6) = 0.22$. Calculate (a) $p(Y < 3.3)$, (b) $p(3.8 < Y < 5.1)$,
(c) $p(|Y - 4.5| > 0.8)$.

6 The crv Z is normally distributed as $Z = N(\mu, 4)$, and
$p(Z > 15.5) = 0.15$.
Calculate (a) μ, (b) $p(Z < 9.6)$, (c) $p(10.4 < Z < 14.2)$,
(d) $p(14.2 < Z < 15.8)$.

7 The crv X has a normal distribution such that $X = N(\mu, \sigma^2)$, and
$p(X > 12.5) = 0.14$, $p(X < 8.1) = 0.18$.

(a) Calculate μ and σ^2.   (b) Calculate $p(|X - \mu| < 4.6)$.

8 The masses of oranges (M g) sold in a local shop are normally distributed
according to $M = N(215, 26.4)$. They are sold in bags containing 10
fruits. Find:

(a) p(the mass of a randomly selected orange > 225 g),
(b) p(the probability that all the fruits in a randomly selected bag each
have a mass less than 220 g),
(c) how many oranges (to the nearest whole number) from a randomly
selected bag you would expect to have masses less than 212 g.

9 The diameters (D cm) of a large batch of ball bearings are found to be
normally distributed as $D = N(\mu, \sigma^2)$. Of this batch, 8% of the bearings
had diameters greater than 1.004 cm and 90% had diameters greater than
0.997 cm.

(a) Find μ and σ^2. Give your answer to μ to five decimal places.
(b) Calculate the proportion of ball bearings whose diameters were greater than 1.002 cm.

10 A circular saw is used to cut metal pipes into sections with an average length of 2.45 m. After heavy use for many months, it is found that the mean length of the pipes remains at 2.45 m, but that 15% of the pipes are rejected as being too long. This implies that they are longer than 2.48 m. The rods are also rejected if they are less than 2.41 m long. Assuming that the lengths are normally distributed, what proportion of rods would you expect to be rejected as being too short?

11 If $X = N(100, 16)$, find the value of k in each of the following.
(a) $p(X > k) = 0.2266$ (b) $p(X > k) = 0.12$
(c) $p(X > k) = 0.9332$ (d) $p(X > k) = 0.6915$
(e) $p(X > k) = 0.9599$

12 If $X = N(0, 4)$, find the value of k in each of the following.
(a) $p(X < k) = 0.0668$ (b) $p(X < k) = 0.4013$
(c) $p(X < k) = 0.1841$ (d) $p(X < k) = 0.8770$
(e) $p(X < k) = 0.6368$

4.3   Normal approximations to the binomial and Poisson distributions

Basic

1 In each part of this question, use the normal approximation to the binomial to calculate the required probabilities. Do not use a continuity correction in this question.

(a) $X = B(20, \frac{1}{2})$: (i) $p(X < 14)$ (ii) $p(X > 8)$
(b) $X = B(30, 0.6)$: (i) $p(X > 20)$ (ii) $p(X < 14)$
(c) $X = B(100, 0.7)$: (i) $p(X < 80)$ (ii) $p(60 < X < 75)$
(d) $X = B(200, 0.4)$: (i) $p(X < 65)$ (ii) $p(X > 84)$
(e) $X = B(140, 0.6)$: (i) $p(X < 91)$ (ii) $p(|X - 84| < 10)$

2 The crv X is defined as $X = B(100, 0.4)$. Use a normal approximation, with continuity correction, to find the following probabilities.

(a) $p(X \geq 45)$ (b) $p(X < 43)$ (c) $p(X \leq 36)$
(d) $p(X \geq 38)$ (e) $p(30 \leq X < 40)$ (f) $p(42 \leq X \leq 50)$

3 In a biscuit factory, 8% of the biscuits are broken during the manufacturing process. A sample of 500 biscuits is taken. Use an

appropriate normal approximation to find the probabilities of the
following numbers of broken biscuits in the sample:

(a) more than 45,   (b) less than 36,
(c) between 37 and 42, inclusive,   (d) at least 37.

*4 An unbiased die is thrown 1800 times and the number of sixes seen is
recorded. Use a normal approximation to determine the probability that
the number of sixes was (a) more than 312, (b) between 298 and 303,
inclusive, (c) less than 290, (d) 303.

5 A company that manufactures cricket balls finds that 4% of the balls are
defective. Find, using an appropriate normal approximation, the
probability that the number of defective balls in a sample of 200 is
(a) more than 8, (b) 7, (c) at least 7, but less than 9.

6 A small, uniform, unbiased cube has one of its faces painted red, two
painted yellow, and the remainder green.
The cube is rolled 150 times and a record is kept of the number of times
a yellow face is uppermost. Use a normal approximation to find the
probability that the number of times a yellow face is recorded is (a) 52,
(b) more than 45, (c) less than 54, but at least 47, (d) at least 58.
What is the probability that a green face is seen more than 90 times?

7 At a school fair, the spinner
shown is used to raise money.
The angles of the four segments
are 180°, 90°, 60° and 30°, and
they are painted red, white, blue
and green respectively. When spun,
the arrow is equally likely to stop at
any point round the circle. A prize
of £1 is paid if the arrow stops
above the green sector.

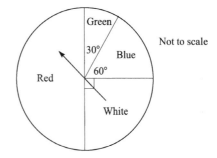

(a) Use a normal approximation to
find the probability that at least
£50 is paid out if the spinner is
used 480 times during the fair.
(b) Use a similar approximation to find the probability that the spinner
stops over the white sector more than 130 times during the fair.

8 Find the probabilities listed below, using a normal approximation with a
continuity correction.

(a) For $X = \text{Po}(30)$, find (i) $p(X \leq 37)$, (ii) $p(X \geq 20)$, (iii) $p(X < 25)$,
(iv) $p(X > 38)$, (v) $p(X > 27)$.
(b) For $X = \text{Po}(54)$, find (i) $p(X \geq 39)$, (ii) $p(47 < X < 68)$,
(iii) $p(X > 70)$, (iv) $p(|X - 54| < 12)$, (v) $p(|X - 54| > 7)$.

9 In a particular mathematics class, the number of times that a pupil puts up a hand to ask or answer a question, per period, follows a Poisson distribution with mean 25.

(a) Find, using a normal approximation, the probability that, in any one period, a hand is put up (i) less than 16 times, (ii) at least 30 times, (iii) at least 19, but not more than 28 times.

(b) The class has six mathematics periods each week. In an eight-week term, in how many periods (to the nearest whole number) would you expect to see a hand raised exactly 30 times?

10 If $X = \text{Po}(40)$, use a normal approximation to find the probability that:

(a) $p(X = 51)$ (b) $p(X = 26)$

Find also the probability that X is greater than or equal to 29 but less than 48.

11 Calculate the following probabilities using (i) the Poisson formula, (ii) a normal approximation.

(a) $X = \text{Po}(40)$; find $p(X = 35)$ (b) $X = \text{Po}(25)$; find $p(X = 20)$

12 Each morning, between 09.00 and 12.00, the secretary of a large school receives on average 50 telephone calls. The frequency of these calls follows a Poisson distribution. Using (a) the Poisson formula, (b) a normal approximation, find the probability that, on a randomly chosen morning, 58 calls are received.

Intermediate

1 All 540 pupils in a school sat two general knowledge tests. The first consisted of 60 questions, each of which required a true-false answer. The test was set by a crusty old statistician who made the questions so obscure that all the pupils had to guess the answers. One mark was given for each correct answer, and the pass mark was set at 28 marks. Show, using a normal approximation, that you would expect about 400 pupils to pass. The second test was set by the same statistician, and was equally difficult, making all the pupils guess again. For this test, pupils had to choose from three possible answers to each question, and there were again 60 questions. The pass mark was 18 marks. Using the same approximation method, find the number of pupils expected to pass.

2 At a university, 240 students are about to take one of their final history examinations. From past records, 5% of the students will score a first grade in this examination. Use a normal approximation to find the probability that the number scoring a first grade is (a) 16, (b) less than 10, (c) 20 or 21.

3 In a large school, there is, in the summer term, a choice of two major games, cricket and tennis. Every pupil is expected to play (however infrequently!) one of these games, and, over several years, the average proportion of pupils choosing tennis is found to be $\frac{25}{42}$. In a particular summer, there are 168 pupils in the sixth form.

(a) Using the above proportions, how many sixth formers would you expect to opt for tennis?

(b) Use a normal approximation to find the probabilities that the number of sixth formers choosing tennis that term is (i) less than 90, (ii) at least 112, (iii) 95.

*4 Joe has difficulty with his spelling. His English teacher checks his essays, and finds that, on average, he makes two mistakes in every 10 lines. A colleague who teaches statistics suggests that the frequency of the mistakes probably follows a Poisson distribution. For English coursework, Joe has to write an essay which, when finished, covers 210 lines.

(a) Use first the Poisson formula and then a normal approximation to it to find the probability that Joe has made 48 mistakes in his essay.

(b) Also find, using the normal approximation, the probability that the number of mistakes is (i) less than 35, (ii) at least 52.

5 A car manufacturer finds that the plastic headlamp lenses supplied by an outside contractor have, on average, 3% of lenses defective.
Use the normal approximation to the binomial distribution to calculate an estimate of the probability of having

(a) 14 defective lenses in a batch of 300,

(b) more than 5 defective lenses in a batch of 100.

6 A supermarket has a promotion to sell one of its 'own name' brands of washing powder. Customers are given a free sample and invited to fill in and post back a small questionnaire about other brands sold in the shop, comparing them with the 'own name' brand. A check on returned questionnaires shows that 42% of the customers prefer the 'own name' brand to any other.
Use the binomial formula and the normal approximation to it to calculate the probability that, out of a random group of 20 customers interviewed after the promotion,

(a) 10 prefer the supermarket's 'own name' brand,

(b) between 8 and 10 inclusive choose the 'own name' brand.

7 Two Poisson variables are defined below. Calculate the listed probabilities using (i) the Poisson formula, (ii) a normal approximation.

(a) $X = \text{Po}(28.5)$; calculate $\text{p}(X = 31)$

(b) $X = \text{Po}(21.6)$; calculate $\text{p}(X = 18)$

8   Use the normal approximation to the Poisson distribution to calculate the following probabilities for the random variable $X = \text{Po}(25.5)$.

(a) (i) $p(X = 29)$ (ii) $p(20 < X \le 28)$ (iii) $p(|X - 25.5| < 7.5)$

(b) (i) $p(|X - 25.5| > 5.5)$ (ii) $p(|X - 25.5| < 1.5)$

9 X is a random variable with a Poisson distribution such that $X = \text{Po}(30)$. Calculate (i) using the Poisson formula, (ii) using a normal approximation:

(a) $p(X = 33)$ (b) $p(X = 24)$

In each case calculate the percentage difference between (i) and (ii), taking the answer to (i) as correct.

10 The random variable X has a binomial distribution such that $X = \text{B}(120, 0.04)$.

(a) Find the following probabilities, using the binomial formula.
     (i) $p(X = 8)$ (ii) $p(X = 9)$ (iii) $p(X = 10)$
(b) Calculate the same probabilities, using, first, the Poisson and, second, the normal approximations to the binomial.

11 A cubical die is biased so that $p(1) = p(6) = 0.4$, the other scores being equally likely. The die is rolled 50 times. Calculate the following probabilities, using the required method:

(a) $p(30$ 'ones' are recorded), binomial formula,
(b) p(four fives are seen), normal approximation to the binomial.

12 Landings of aircraft at London airport at a peak period follow a Poisson distribution and are separated, on average, by 90 seconds. A record is made of the number of landings in an hour. Use a normal approximation to calculate the following probabilities:

(a) p(less than 50 aircraft land in the hour),
(b) p(between 30 and 40 inclusive land in the hour).

13 Ripe apples drop off a tree at intervals which are, on average, 75 seconds, and which we assume follow a Poisson distribution. Isaac, keeping well out of the way, observes the tree over a period of 1 hour and counts the apples which drop.

(a) Use both the Poisson formula and the normal approximation to the Poisson distribution to calculate the probability of Isaac seeing 35 apples fall.
(b) Use the normal approximation to calculate the probability that more than 60 apples drop in the hour.

14 A weaver, sitting at her loom, finds that, on average there are two small faults (usually caused by imperfections in the wool being used) in every 20 rows of the rug she is making. We assume that the spacing of the faults follows a Poisson distribution. In the finished rug, there are 420 rows. Using in each case both the Poisson formula and the normal approximation, calculate the probability of (a) 42 faults, (b) 36 or 37 faults in the finished rug.

15 Probabilities for values of the random variable X, defined as B(200, 0.08), are to be calculated using a Poisson distribution as an approximation. Find:

(a) p($X = 20$)   (b) p($X = 10$ or 11)

Advanced

1 A cubical die is biased so that the probabilities of obtaining the six scores are as follows.

Score	1	2	3	4	5	6
Probability	$\frac{1}{12}$	$\frac{1}{6}$	$\frac{1}{4}$	$\frac{1}{4}$	$\frac{1}{6}$	$\frac{1}{12}$

(a) The die is rolled 200 times. Find the following probabilities, using the methods required:

  (i) p(more than 60 fours are seen) – use the normal approximation to the binomial,
  (ii) p(exactly 44 threes are seen) – use both the binomial formula and a normal approximation,
  (iii) p(3 or 4 is seen more than 90 times) – use a normal approximation.

(b) The die is rolled 120 times. Find the following probabilities, using the requested methods:

  (i) p(a five is seen 18 times) – use the binomial formula and the normal approximation to it,
  (ii) p(more than 25 twos are seen) – use a normal approximation.

*2 An electronics factory makes rechargeable back-up batteries for video recorders. On average, it is found that 2% of these batteries are not up to standard.
Using (a) the binomial formula, (b) a normal approximation, (c) a Poisson approximation, calculate the probability that three or more batteries from a batch of 100 are below standard. Comment on the closeness of the approximations.

3 A car manufacturer discovers a serious defect in the steering linkage of one of its models, the defect occurring in 25% of the large number already sold. All the cars sold during a certain period are recalled to the selling agents for modification.

   (a) A large dealer sold 84 of these cars and they are all brought in for checking. Use a suitable approximation to find the probability that: (i) less than 15 have the defect, (ii) between 18 and 28 inclusive are defective.

   (b) If more than 28%, to the nearest whole number, of the cars sold by any garage have to be modified, extra compensation will be paid by the factory. For a garage that sold 64 cars, what is the probability that it will qualify for compensation, if all 64 cars are brought in?

4 A cubical die is biased so that $p(6) = \frac{2}{5}$, all the other scores being equally likely. Use both the binomial formula and the normal approximation to the binomial to find the following probabilities, if the die is rolled 100 times:

(a) 30 sixes are seen, (b) 48 sixes are seen, (c) 18 'ones' are observed, (d) six 'ones' appear.

Which is more likely to occur, 25 sixes or 20 'ones'?

5 The opening hours for a doctor's surgery are from 09.00 to 12.00 daily from Monday to Friday. Patients arrive randomly and a check by a secretary shows that they arrive at an average rate of five per hour.

   (a) Assuming arrivals can be modelled by a Poisson variable X, calculate the following probabilities, X being described as $X = $ Po (the average number of patients arriving per hour).
     (i) $p(X = 7)$ (ii) $p(X < 7)$ (iii) $p(2 < X \le 8)$

   (b) Find the probability that 20 patients arrive on a randomly chosen day.

   (c) In a randomly observed week of five days, what is the probability that the number of patients attending is between 65 and 90 inclusive? Use a suitable approximation for this result.

6 Outside a mainline railway station is a taxi-rank. Taxis return to the rank after journeys so that the average number arriving in a period of T minutes has a Poisson distribution with parameter $\frac{1}{3}T$.

   (a) Find the probability that fewer than five cabs arrive in a randomly chosen half-hour period.

   (b) Over a period of two hours, selected at random, find the probability that the number of taxis arriving at the rank is (i) more than 45, (ii) between 45 and 55 inclusive.

7 A statistician with a keen interest in bird life has a large bird table in a prominent but safe place outside his sitting-room window. He notices that one of the sparrows which visit the table regularly has a red ring on one of its legs, so he is able to record visits to the table. After a time, he finds that the average interval between visits is three minutes, and that the number of visits per unit time follows a Poisson distribution during the daylight hours of observation.

(a) Calculate the probability that, in a randomly selected period of 30 minutes, the bird makes (i) five visits, (ii) fewer than four visits to the table.

(b) Calculate the probability that, when the table is kept under continuous observation for two hours, the number of visits is (i) more than 50, (ii) between 40 and 50 inclusive.

8 The proportion of teenagers in the UK population who are short-sighted may be taken to be 30%.

(i) Use a binomial distribution to calculate the probability that at least 3 out of a random sample of 12 teenagers will be short-sighted.

(ii) A family in which there are 3 teenagers is chosen at random from all such families in the UK. Given a reason why the use of a binomial distribution to calculate the probability of at least two of the teenagers in the family being short-sighted might not be appropriate.

(iii) A random sample of 1000 teenagers is taken. Calculate the probability that at least 340 will be shortsighted.

[UCLES]

9 It is given that $X = B(16, 0.5)$.

(i) Calculate $P(X = 4)$ directly, using the binomial distribution.

(ii) Calculate $P(X = 4)$ using a Normal approximation

[UCLES]

Revision

1 Calculate the probabilities requested below, using the required methods.

Distribution	Calculate	Methods
(a) $B(60, \frac{1}{2})$	$p(X = 40)$	(i) binomial formula
		(ii) normal approximation
(b) $B(32, 0.25)$	$p(X < 3)$	(i) binomial formula
		(ii) normal approximation
(c) $B(100, 0.1)$	$p(X = 15)$	(i) binomial formula
		(ii) Poisson approximation
(d) $Po(22)$	$p(X = 25)$	(i) Poisson formula
		(ii) normal approximation.

2 Find the normal approximations to the following binomial probabilities.

(a) For $X = B(12, 0.45)$, find (i) $p(X > 4)$, (ii) $p(2 < X < 6)$, (iii) $p(X > 9)$.

(b) For $X = B(36, \frac{2}{3})$, find (i) $p(X < 19)$, (ii) $p(X > 20)$,
(iii) $p(|X - 24| < 5)$, (iv) $p(|X - 24| > 2)$.

3 Several binomial variables are defined below. Use either the binomial formula or the most suitable approximation to calculate the listed probabilities.

(a) $X = B(15, 0.55)$:   (i) $p(X = 10)$ (ii) $p(X < 11)$ (iii) $p(X < 2)$
(b) $X = B(20, 0.48)$:   (i) $p(X = 19)$ (ii) $p(X = 15)$ (iii) $p(X < 15)$
(c) $X = B(4000, 0.01)$: (i) $p(X < 30)$ (ii) $p(X > 45)$

4 Use either the Poisson formula, or, where it is more suitable, the normal approximation to find the following Poisson probabilities.

(a) $X = Po(24)$:   (i) $p(X = 20)$ (ii) $p(X < 15)$
(b) $X = Po(40)$:   (i) $p(X = 30)$ (ii) $p(X > 30)$

5 A university student takes a temporary job checking marked examination papers. He finds that he can check, on average, 18 papers in an hour. He works a six-hour day, and it is assumed that the number of papers checked per hour follows a Poisson distribution. Find the following probabilities, using a normal approximation if appropriate, where X is the crv, 'the number of papers checked in a day'.

(a) $p(X = 90)$ (b) $p(X > 125)$ (c) $p(88 < X < 130)$

*6 A farmer uses a small explosive device as a bird scarer in his apple orchard. The device is programmed to go off at random intervals, between 06.00 and 19.00, so that there are four explosions in a 30-minute period.

(a) Find the probability that there are, in a randomly chosen period of one hour,
(i) three explosions, (ii) not more than four explosions.
(b) Use the normal approximation to the Poisson distribution to find the probability that the neighbours complain of more than 125 explosions in the course of a randomly selected day.

7 A coin is biased so that $p(\text{heads}) = \frac{3}{4}$.

(a) The coin is spun 20 times. Calculate the probability that
(i) 18 heads are recorded, (ii) ten tails are recorded.
(b) The coin is now spun 200 times. Calculate, using a normal approximation, the probability that (i) a head appears more than 140 times, (ii) a tail appears 65 times.

8 A wholesale egg-farmer supplies a supermarket chain. On a particular day, she delivers a consignment of 10 000 large eggs to the distribution centre for a northern county. The quality control section of the supermarket know that, on average, 3% of eggs are broken on arrival. Use a suitable approximation to determine the probability that:

(a) in a sample of 1000 eggs, fewer than 20 are broken,
(b) in the complete batch of 10 000, fewer than 275 are broken.

9 The random variables V and W have Poisson distributions defined as $V = \text{Po}(10)$, $W = \text{Po}(35)$. Compare each of the following Poisson probabilities with the result obtained from the corresponding normal approximation, and comment on the accuracy of the approximation.

(a) $\text{p}(V = 3)$ (b) $\text{p}(V < 6)$ (c) $\text{p}(W = 37)$ (d) $\text{p}(W < 20)$

10 The random variables X and Y are defined as $X = \text{B}(10, 0.52)$, $Y = \text{B}(500, 0.6)$. Calculate the following exact binomial probabilities and compare them with the values obtained from the normal approximations.

(a) $\text{p}(X = 6)$ (b) $\text{p}(X < 6)$ (c) $\text{p}(Y = 12)$ (d) $\text{p}(Y < 12)$

11 Three random variables, X_1, X_2 and X_3 are defined as $X_1 = \text{B}(75, 0.05)$, $X_2 = \text{B}(16, 0.4)$, $X_3 = \text{Po}(35)$.
Choose a suitable approximation to calculate the following probabilities.

(a) $\text{p}(X_1 = 12)$ (b) $\text{p}(X_2 > 7)$ (c) $\text{p}(X_3 < 30)$.

12 A fast bowler (cricket) bowls, on average, a 'yorker' (near the batsman's toes) once an over. During a test match innings, this bowler bowls 36 overs, each of six balls.
Use the binomial formula, or a suitable approximation to it, to find the probability that the bowler, in this innings, bowled:

(a) 6 'yorkers',   (b) between 8 and 10 'yorkers' inclusive,
(c) more than 9 'yorkers'.

4.4   Linear combinations of variables

Basic

1 Calculate the probabilities listed below, where two normal variables are combined in various ways and $X = \text{N}(20, 4)$, $Y = \text{N}(10, 9)$.

(a)   (i) $\text{p}(X + Y < 31.5)$ (ii) $\text{p}(X + Y < 35)$
       (iii) $\text{p}(X + Y > 33)$ (iv) $\text{p}(X + Y > 34)$
        (v) $\text{p}(X + Y > 28)$ (vi) $\text{p}(X + Y > 26.5)$
       (vii) $\text{p}(X + Y < 26)$ (viii) $\text{p}(X + Y < 27.5)$.
(b)   (i) $\text{p}(X - Y < 12)$ (ii) $\text{p}(X - Y > 11.4)$
       (iii) $\text{p}(X - Y < 8.2)$ (iv) $\text{p}(X - Y > 7.6)$
        (v) $\text{p}(8.3 < X - Y < 12.3)$ (vi) $\text{p}(5.5 < X - Y < 7.2)$
       (vii) $\text{p}(|X - Y - 10| < 2.6)$ (viii) $\text{p}(|X - Y - 10| > 1.8)$

(c)   (i) p(Y − X < −5.2)   (ii) p(Y − X < −8)
    (iii) p(Y − X > −6.3)   (iv) p(Y − X > −9.1)
    (v) p(Y − X > −12.6)   (vi) p(Y − X > −15.7)
    (vii) p(−11.6 < Y − X < −8.5)   (viii) p(|Y − X + 10| < 2.8)

2   A and B are crv's defined as $A = \mathrm{N}(5, 0.09)$, $B = \mathrm{N}(10, 0.25)$.
    Write down the distributions of the following combinations:

    (a)   three independent observations of A, added to five of B,
    (b)   ten independent observations of A, minus two of B,
    (c) $6A$, (d) $4B$, (e) $3A + 6B$, (f) $B − A$, (g) $\frac{1}{2}(A − B)$.

3   Three crv's A, B and C are defined as $A = \mathrm{N}(12, 4)$, $B = \mathrm{N}(20, 9)$,
    $C = \mathrm{N}(15, 2)$. Find the distributions of the following combinations.

    (a) $A + 2B$ (b) $3B + 4C$ (c) $2A − C$
    (d) $B − 3A$ (e) $\frac{1}{2}(2C − B)$ (f) $A + B + C$
    (g) $2A + 3B − 4C$ (h) $2(A + B) + 3C$ (i) $5A − B − 2C$
    (j) $\frac{1}{2}(A + B) + \frac{1}{3}C$

*4   The mass of a loaf of bread produced at the Bakefast bakery is normally
    distributed with mean $800\,\mathrm{g}$ and variance $64\,\mathrm{g}^2$.
    Calculate the probability that:

    (a)   the mass of a loaf, selected at random is greater than $815\,\mathrm{g}$,
    (b)   the total mass of the eight loaves on a tray is less than $6360\,\mathrm{g}$.

5   Tom catches a bus to school, and he arrives promptly at 08.00 at the bus
    stop each morning. The time that he has to wait at the stop is normally
    distributed with mean 5 minutes and variance 2 minutes. What is the
    probability that he will have to wait at the stop for more than 7 minutes
    on any one day? Find the probability that his wait will be between 4 and
    6 minutes.

6   The crv's X and Y are defined in the table:

	μ	σ
X	25	5
Y	30	4

    (a)   Write down the distributions of the following combinations of
        X and Y.

    (i) $X + Y$ (ii) $X − Y$ (iii) $Y − X$ (iv) $2X + Y$
    (v) $4X − 3Y$ (vi) $2X + \frac{1}{2}Y$

    (b)   Calculate:

    (i) p(X − Y < −8)   (ii) p(|X − Y| < 3)
    (iii) p(|2X + Y − 80| < 10)   (iv) p(2X + \frac{1}{2}Y > 80)

7 X and Y are crv's defined as $X = N(25, \sigma^2)$, $Y = N(\mu, 9)$. You are given that $p(X < 27) = 0.8413$, $p(Y > 36) = 0.0228$. Calculate:

(a) μ, σ^2 (b) $p(X + 2Y < 76)$

8 X and Y are crv's defined as $X = N(\mu, \sigma^2)$, $Y = N(4\mu, 4\sigma^2)$. You are given that $Y - X = N(30, 20)$. Find μ and σ^2 and hence the distributions of X and Y.

9 X is the crv such that $X = N(k, k^2)$ and $p(X < 7.5) = 0.6915$. Calculate the value of k.

10 The mass B g of a pair of binoculars is normally distributed as $B = N(660, 50)$, and the mass C g of its case is also a normal variable, $C = N(85, 16)$.
Find the probability that:

(a) a pair of binoculars has a mass between 655 g and 670 g,
(b) the total mass of two pairs of binoculars and their cases is more than 1510 g.

11 Two candles, one large and one small, have burning times, in hours, which are normally distributed as $L = N(16, \frac{1}{4})$, and $S = N(6, \frac{1}{16})$.

(a) What is the probability that:

(i) a large candle burns for less than 14.8 hours,
(ii) a small candle burns for at least $6\frac{1}{2}$ hours?

(b) A large candle is lit. At the instant it is finished, a second large candle is lit. This process is continued with two small candles. What is the distribution for the total burning time of these four candles?

12 In a one-day international cricket match, the first team to bat scored runs at the rate of 5.2 per over. This rate is assumed to be normally distributed with a variance of 4.4.

(a) What is the probability that, in a randomly selected over, the team scored more than 8 runs?
(b) In a randomly selected group of five consecutive overs, what is the probability that this team scored exactly 15 runs?

Intermediate

1   The crv's X and Y are defined as $X = N(50, 16)$, $Y = N(30, 25)$.
Calculate:

(a) $p(2X < 105)$ (b) $p(90 < 2X < 116)$
(c) $p(0.5X < 22)$ (d) $p(X + Y < 88)$
(e) $p(X > Y + 6)$ (f) $p(190 < 2X + 3Y < 220)$

*2 If P and Q are crv's, defined as $P = N(40, 1.2^2)$, $Q = N(20, 2.3^2)$, calculate:

(a) $p(P - Q > 25)$ (b) $p(130 < \{2P + 3Q\} < 145)$
(c) $p(Q > P - 15)$ *(d) $p(2Q > P)$
*(e) $p(P + Q - 58 < 0)$

3 The crv's A and B are defined as $A = N(2.5, 0.098)$, $B = N(3.7, 0.06)$. Find the following probabilities.

(a) $p(A + B < 6.35)$ (b) $p(3A - 4B > -7.8)$
(c) $p(4A - 4B > A - 2B)$

4 The mass of cod-liver oil, M mg, in a capsule is normally distributed as $M = N(500, 25)$. The transparent outer covering has a mass, C mg, which is also normally distributed as $C = N(210, 20)$. The capsules are sold in plastic bottles containing 150 capsules.
Find the probability that:

(a) a single capsule, chosen at random, contains less than 496 mg of oil,
(b) a single, randomly chosen capsule has a total mass of more than 720 mg,
(c) the contents of a full bottle have a mass of more than 106.4 g.

5 Twenty-six clubs play in a national hockey league, and they all play each other twice during a season, once at home and once away. In a particular year, the winning club had goal-scoring records per match for home (H) and away (A) matches as follows: $H = N(3, 0.25)$, $A = N(1.5, 0.2)$.

(a) Find the probability that, in a randomly selected game,
    (i) more than two goals were scored at home,
    (ii) more than two goals were scored in an away game.
(b) At the end of the season, what is the probability that this team has scored more than 120 goals?
(c) In how many home games, to the nearest whole number, would you expect this team to score more than four goals?

6 The crv's X and Y are defined as $X = N(\mu, \sigma^2)$, and $Y = N(100, 64)$. You are given that $p(X > 43) = 0.8413$ and $p(X < 60.5) = 0.9332$. Calculate:

(a) μ and σ (b) $p(2X + Y < 225)$

7 Ruled writing paper has line spacing which is termed Feint (F) or Narrow Feint (F_N). The spacings of both types, in millimetres, are normally distributed as $F = N(8, 10^{-4})$, and $F_N = N(6, 8.1 \times 10^{-5})$.

(a) Find (i) $p(7.985 < F < 8.015)$, (ii) $p(2F + 3F_N > 33.93)$.
(b) What is the probability that the total distance between the first and last lines of eleven consecutive Feint lines is less than 80.02 mm?

8 Two coins are each tossed 900 times and the number of heads recorded. The first coin is fair (F), but the second is biased (B) so that p(heads) $= \frac{1}{3}$. If F and B represent the respective normal approximations to these two binomial distributions, find:

(a) p($F + B > 475$)   (b) p($431 < F + B < 468$)

9 A4 notepaper is cut to a length, L mm, and width, W mm, which are normally distributed as $L = \text{N}(297, 0.04)$ and $W = \text{N}(210, 0.03)$.

(a) Write down the distribution of the perimeter, P mm, of a sheet of A4 notepaper.

(b) Calculate the probability that this perimeter is (i) less than 1014.16 mm, (ii) between 1013.92 mm and 1014.15 mm.

10 X and Y are crv's normally distributed as $X = \text{N}(0.5, 0.002), Y = \text{N}(0.7, 0.004)$.

(a) Write down the following distributions.

(i) $4X + 10Y$ (ii) $5X - 5Y$ (iii) $5(X - Y)$ (iv) $\frac{1}{5}X + \frac{1}{7}Y$

(b) Find the following probabilities.

(i) p($4X + 10Y < 9.55$)   (ii) p($5X > 5Y$)   (iii) p($\frac{1}{5}X + \frac{1}{7}Y > 0.167$)

11 The mass of sweetcorn, S g, in a small tin is normally distributed as $S = \text{N}(330, 15)$. The mass in a large tin, L g, is also normally distributed as $L = 3S$.

(a) What is the probability that the mass of sweetcorn in a small tin is within 10 g of the mean mass?

(b) Write down the distribution of L.

(c) The mean mass of L is μ_L. M_1 and M_2 are two masses (in grams) such that $\mu_L - M_1 = M_2 - \mu_L$ and p($L < M_2$) $= 0.8$. Calculate, to the nearest whole number, the values of M_1 and M_2.

*12 The average rainfall, R inches, at a seaside resort, in the West Country in September is described by the normal variable $R = \text{N}(3, 1)$.

(a) Calculate the probability of there being, in a randomly selected September,

(i) more than 4.2 inches of rain in the whole month,

(ii) more than a quarter of an inch of rain on a randomly chosen day.

(b) On how many days, to the nearest whole number, would you expect more than 0.3 inches of rain?

13 An athlete runs a large number of 800 m races in a season, and her coach, who happens to know some statistics, finds that her average time is a normal variable, T seconds, described as $T = \text{N}(128, 12)$.

(a) Find the probability that

(i) in a randomly selected race, her time is less than 124.5 seconds,
(ii) her total time, for four randomly chosen races, is more than 520 seconds.

(b) If she took part in 25 races during the season, in how many of these, to the nearest whole number, would you expect her to record a time between 124 and 130 seconds?

14 A company, for its annual Christmas dinner, asks an artistic employee to draw on each menu the company logo, which is an equilateral triangle of side 3 cm. The company director who is pernickity about the logo's accuracy, measures all the triangles and finds the sides to have a length L cm, which is a normal variable such that $L = N(3, 0.01)$. (She then accepts the drawings!)
Find the probability that:

(a) one side of a randomly selected triangle is less than 2.8 cm long,
(b) the perimeter of a triangle chosen at random is more than 9.3 cm.

15 Three crv's are defined as $X = N(2k, k^2)$, $Y = N(3k, 4k^2)$, $Z = N(5k, 9k^2)$, where k is a positive real number. Calculate:

(a) $p(X < 2.6k)$ (b) $p(Y > 1.5k)$ (c) $p(Y + Z > 13k)$
(d) $p(3k < 3X < 12k)$

Advanced

1 Dried fruit is sold in a supermarket in packets whose mass is normally distributed with mean 500 g and standard deviation 2 g.

(a) What is the probability that a single, randomly chosen packet weighs less than 498.5 g?
(b) The packets are delivered to the supermarket in boxes of 20 packets. What is the probability that the contents of the packets in a randomly selected full box weigh more than 10 009 g?
(c) As a special purchase, the supermarket buys in a batch of jumbo boxes, each of which contains the contents of four of the 500 g packets. Find the probability that the contents of a randomly selected jumbo box have a mass greater than 2020 g.

*2 X and Y are crv's such that $X = N(\mu, \sigma^2)$, $Y = N\left(\dfrac{4\mu}{3}, \dfrac{16\sigma^2}{9}\right)$. You are given that $p(X + Y < 40) = 0.8413$, and $p(X + Y < 45) = 0.9772$. Find the values of μ and σ and hence find $p(Y < X)$.

3 A bar of milk chocolate has a mass which is normally distributed with mean 100 g and standard deviation 1 g. Its wrapping has a mass which is also normally distributed with mean 2 g and standard deviation 0.4 g.

(a) What is the probability that:

(i) the mass of a randomly chosen bar will be greater than 102.4 g,

(ii) the total mass of a randomly chosen bar and its wrapping is more than 104 g?

(b) The bars are packed in boxes holding 10 bars. What is the probability that the mass of the chocolate plus wrapping in a randomly selected box is greater than 1025 g?

4 A packet of Breathfresh chewing gum contains five sticks of gum. The mass, S g, of a stick is normally distributed as $S = N(2, 0.09)$, and it is wrapped in foil and paper whose mass, P g, is also normally distributed as $P = N(0.5, 0.0064)$. The five sticks plus wrapping are enclosed in a plastic case, the mass, C g, of which is also normally distributed as $C = N(0.84, 0.0196)$.

If each selection is made at random, calculate the probability that:

(a) the mass of an unwrapped stick is less than 1.75 g,

(b) the mass of a single stick and its wrapping is more than 2.85 g,

(c) the total mass of a packet of five sticks and all the wrapping is between 11.5 and 14.2 g.

5 A firm which makes model aircraft paints two different logos, one circular and one triangular, on each model. The volumes of paint required, in millilitres, are normally distributed such that, for the circular logo, $C = N(5, 0.16)$ and, for the triangular logo, $T = N(4, 0.09)$.

(a) Find the probability that:

(i) for a single circular logo, the volume of paint used is more than 5.4 ml,

(ii) for five triangular logos, the volume of paint used is less than 21.5 ml.

(b) If 100 of each logo are painted, find the probability that the total volume of paint used is less than 890 ml.

6 The lengths, L cm, of the legs for a cheap, flat-pack garden bench are normally distributed as $L = N(58, 2.25)$.

(a) Find the probability that:

(i) a single leg has a length greater than 60 cm,

(ii) a set of four legs has a total length of more than 240 cm, these selections being made at random.

(b) The factory cuts two legs from pieces of wood whose lengths, L_1 cm are also normally distributed so that $L_1 = N(120, 4)$. What is the probability that the combined lengths of two legs exceed that of a single piece of this wood?

7 Large beer cans contain a volume of beer which is normally distributed with mean 500 ml and standard deviation 3.3 ml, while the volume in small cans is also normally distributed with mean 340 ml and standard deviation 2.4 ml.

(a) If cans are chosen at random from stock for testing, find the probability that:

   (i) a large can contains less than 495 ml of beer,
   (ii) the volume of beer in a small can is between 337 and 342 ml,
   (iii) a dozen cans of each size contain in total more than 10 060 ml of beer,
   (iv) a dozen cans of each size contain in total between 10 050 and 10 070 ml of beer.

(b) One hundred large cans are tested. How many, to the nearest whole number, would you expect to contain more than 497 ml of beer?

(c) An extra large can contains the contents of four large cans. What is the probability that a case of six of these extra large cans contains in total more than 12 040 ml of beer?

8 The independent random variables X and Y are such that

$$X = N(48, 92) \quad \text{and} \quad Y = N(20, 18)$$

(i) Find $P(\frac{3}{4}X + \frac{1}{4}Y > 50)$.
(ii) Two independent observations of Y are denoted by Y_1 and Y_2.

   Find $P(Y_1 + Y_2 > X)$.

[UCLES]

9 Every summer many competitors attempt to climb the British mountains Ben Nevis, Scafell Pike and Snowdon. For each mountain the times, in minutes, taken by the competitors are normally distributed, with means and standard deviations given in the following table.

	Mean	Standard deviation
Ben Nevis	230	40
Scafell Pike	170	25
Snowdon	190	30

Assuming that the times for the three mountains are independent of one another, calculate the probability that

   (i) a single randomly chosen competitor climbs Ben Nevis in less than 180 minutes,
   (ii) the total time taken by a single randomly chosen competitor to climb all three mountains lies between 540 and 660 minutes,
   (iii) a single randomly chosen competitor takes longer to climb Scafell Pike than Snowdon,

(iv) a randomly chosen team of four competitors, walking independently of each other, climbs all three mountains in a total time greater than 2520 minutes. Give reasons to explain why it may be inappropriate to assume the independence of (a) the times for the three mountains, (b) the times of the team members.

[UCLES]

Revision

1 X and Y are crv's which are normally distributed as $X = N(150, 80)$, $Y = N(120, 60)$.

(a) Write down the following distributions:

(i) $3X$
(ii) $(Y_1 + Y_2 + Y_3 + Y_4 + Y_5)$, where these are random observations from the distribution of Y
(iii) $4X - 3Y$ (iv) $\frac{1}{2}X - 2Y$ (v) $5Y - 2X$

(b) Calculate:

(i) $p(3X < 475)$
(ii) $p(570 < \{Y_1 + \cdots + Y_5\} < 600)$
(iii) $p(|4X - 3Y - 240| < 40)$
(iv) $p(\frac{1}{2}X - 2Y > -195)$
(v) $p(|5Y - 2X - 300| > 50)$

2 The crv X is defined as $X = N(\mu, 8)$, and you are given that $p(X < 66.7) = 0.83$. Calculate to three significant figures the value of μ.

3 X is a crv defined as $X = N(2k^2, k^2)$, where k is a positive integer, and it is given that $p(X > 12) = 0.9772$. Calculate the value of k.

4 All the 76 pupils taking GCSE mathematics in a large school take a test which is marked out of 100. The average mark was calculated to be 44% and the standard deviation was found to be 14 marks.

(a) If we assume that the marks were normally distributed, calculate the probability that:

(i) a randomly chosen pupil's score was 56,
(ii) the marks of three randomly selected pupils totalled less than 173.

(b) How many pupils would you expect to score (i) more than 60%, (ii) less than 25%. (Give these answers to the nearest whole number. Use continuity corrections.)

5 A crv X is defined as $X = N(\mu, \sigma^2)$. You are given that $p(X < 50) = 0.9$, and $p(X < 40) = 0.45$.
Calculate, to three significant figures, the values of μ and σ and so find $p(2X < 75)$.

6 X and Y are crv's defined as $X = N(\mu, \sigma^2)$, $Y = N(2\mu, 2\sigma^2)$. You are given that $2X + 3Y = N(80, 88)$. Calculate:

(a) μ and σ^2 (b) $p(4X < 3Y)$ (c) $p(75 < 5X + 2Y < 100)$

*7 A and B are two normally distributed crv's defined as $A = N(\mu, \sigma^2)$ and $B = N\left(4\mu, \dfrac{7\sigma^2}{3}\right)$. You are given that $4B - 3A = N(260, 1390)$.

Find (a) the distributions of A and B, (b) $p(2A + B < 130)$.

8 An agricultural firm buys in two sizes of fertiliser bag, small (S) and large (L). When full, these hold volumes of fertiliser, in litres, which are normally distributed according to $S = N(40, 0.32)$, $L = N(90, 0.42)$. Calculate:

(a) $p(S < 39)$ (b) $p(|L - 90| < 1)$ (c) $p(257 < 2S + 2L < 262)$

9 Four athletes train for a 4×100 m relay race. After many training races, in which they all run in the same order, the average time for each leg, L seconds, is found to be normally distributed according to the following definitions:

$$L_1 = N(11.3, 0.0081), \quad L_2 = N(11.1, 0.01),$$
$$L_3 = N(11.1, 0.0064), \quad L_4 = N(11.2, 0.0121)$$

Assuming all handovers are perfect, i.e. each athlete runs exactly 100 m with the baton, calculate:

(a) $p(L_1 < 11.2)$ (b) $p(L_1 + L_2 < 22.2)$
(c) $p(L_4 > L_3)$ (d) $p(L_1 + L_2 + L_3 + L_4 < 44.4)$

10 The crv's X and Y are defined as $X = N(0.2, 0.001)$, $Y = N(0.08, 0.0002)$.

(a) Write down the following distributions:
   (i) $X + Y$ (ii) $Y - X$ (iii) $2Y - 5X$ (iv) $\frac{1}{10}X + \frac{1}{16}Y$
(b) Calculate the following probabilities.
   (i) $p(X + Y > 0.31)$ (ii) $p(Y > X - 0.08)$
   (iii) $p(2Y - 5X < -0.5)$ (iv) $p(\frac{1}{10}X + \frac{1}{16}Y < 0.032)$

11 A car leasing firm has a policy of selling off its vehicles each year. The average mileage, M miles, on the clock is normally distributed as $M = N(10\,400, 250\,000)$.

(a) For a randomly observed model, calculate the following probabilities.
   (i) $p(M < 9250)$ (ii) $p(|M - 10\,400| < 850)$
(b) Find the probability that, if the mileages of four randomly chosen cars are added, the total will be less than 42 500 miles.

12 The plastic barrel of a felt-tip pen has a diameter, D cm, which is normally distributed as $D = N(1, 0.0004)$.

(a) Find (i) $p(D < 1.015)$,   (ii) $p(|D - 1| < 0.014)$.
(b) A plastic desktop holder is made for this pen. It has a cylindrical hole in its upper surface which contains the pen. The diameter, D_1 cm, of this hole is also normally distributed such that $D_1 = N(1.03, 0.0001)$.

   (i) What is the probability that, if a pen and holder are chosen at random, the diameter of the pen will be larger than the diameter of the hole?

  (ii) What would have to be the average diameter of the hole in the holder, to three significant figures, its variance being unchanged, for this probability to be reduced to 0.02?

5

Correlation and regression

Scatter diagrams, regression lines, correlation coefficients

Regression lines:

$$y \text{ on } x, \quad y - \bar{y} = m(x - \bar{x}), m = \frac{n \sum xy - \sum x \sum y}{n \sum x^2 - (\sum x)^2}$$

$$x \text{ on } y, \quad x - \bar{x} = m'(y - \bar{y}), m' = \frac{n \sum xy - \sum x \sum y}{n \sum y^2 - (\sum y)^2}$$

Product moment correlation coefficient

$$r = \frac{n \sum xy - \sum x \sum y}{\sqrt{n \sum x^2 - (\sum x)^2} \sqrt{n \sum y^2 - (\sum y)^2}}$$

Spearman's rank correlation coefficient,

$$\rho = 1 - \frac{6 \sum d^2}{n(n^2 - 1)}$$

Basic

1   Plot scatter diagrams for each of the following sets of points, find $(\bar{x}, \bar{y})$ and draw a line of best fit through this point. Comment on the degree of correlation in each case.

(a)

x	2	4	6	8	10	12
y	7	9	8	9	13	14

(b)

x	1	2	3	4	5	6	7	8
y	1	3	3.5	2	4	7	5.5	6

(c)

x	1	4	5	5	6	7
y	2	4	6	8	10	12

(d)

x	5	10	15	20	25
y	8	7	4	4	2

(e)

x	2.5	7.5	12.5	15	20	25
y	7	5.5	5	3	2.5	1

2 For each set of data below draw the scatter diagram. Calculate the coordinates of $(\bar{x}, \bar{y})$ and hence draw the regression lines of y on x and x on y. Label the lines clearly.

(a)

x	2	3	4	5	6	7	8	9
y	3	5	9	7	10	13	14	19

(b)

x	2	4	6	8	10	12	14	16
y	20	17	18	13	13	15	9	7

(c)

x	3	5	6.5	9	12	14	16.5	18	20	23
y	4.1	7	6	11	12.9	10.5	17.5	15.7	19.5	19.8

3 Draw a scatter diagram for each of the sets of points below. Calculate $M(\bar{x}, \bar{y})$. Use this to draw the regression line requested and then find the values of x or y as requested.

(a)

x	2	4	6	8	10	12	14	16	18	20
y	4	5	3	7	5	10	7	13	11	15

Draw the regression line of y on x and find y when $x = 15$.

(b)

x	1	2	3	4	5	6	7	8
y	5.5	7.5	6	5	2.5	2	1.5	2

Draw the line x on y and find x when $y = 3.5$.

(c)

x	2	3	4	5	6	7	8	9	10	11
y	18	18	16	12	8	10	10	4	4	4

Draw the line y on x and find y when $x = 4.5$.

(d)

x	5	10	15	20	25	30	35	40
y	57	50	49	38	37	42	31	26

Draw the line x on y and find x when $y = 27.5$.

4 For each set of data given, calculate:

(i) $S_{xy} = \dfrac{\sum xy}{n} - \bar{x}\bar{y}$ (ii) $S_x^2 = \dfrac{\sum x^2}{n} - \bar{x}^2$ (iii) $S_y^2 = \dfrac{\sum y^2}{n} - \bar{y}^2$

(a)

x	2	3	4	5	6	7
y	3	5	7	9	11	13

(b)

x	5	10	15	20	25	30
y	14	13	11	10	8	7

(c)

x	4	8	12	16	20	24
y	21	18	14	11	9	7

(d)

x	2	5	9	11	14	18	20	25	27	30
y	5	6.5	7.2	9	10.8	10.7	12.3	14	15.9	17.2

*5 Find the product moment correlation coefficient (r) for each set of data below and comment on the degree of correlation it implies.

(a)
x	1	2	3	4	5	6
y	7	10	13	16	19	22

*(b)
x	1	2	3	4	5	6
y	5	12	7	15	7	12

(c)
x	1	2	3	4	5	6
y	5	12	2	16	8	7

(d)
x	2	4	6	8	10	12
y	12	10	8	6	4	2

(e)
x	3	6	9	12	15	18	21
y	18	7	14	11	4	6	3

6 From the summarised data given below, calculate:

(i) the product moment correlation coefficient,
(ii) the regression lines of y on x and x on y.

(a) $\sum x = 25$, $\sum x^2 = 135$, $\sum y = 53$, $\sum y^2 = 777$, $\bar{x} = 10.6$
$\bar{y} = 10.6$, $\sum xy = 311$

(b) $\sum x = 31$, $\sum x^2 = 229$, $\sum y = 55$, $\sum y^2 = 775$, $\bar{x} = 6.2$,
$\bar{y} = 11$, $\sum xy = 294$

(c) $\sum x = 67$, $\sum x^2 = 915$, $\sum y = 71$, $\sum y^2 = 1159$, $\sum xy = 1022$,
$n = 6$

7 From the equations of the regression lines given, calculate the product moment correlation coefficient.

(a) $y = 4 + 0.2x$, $x = 3 + 4y$
(b) $y = 0.3 - 1.4x$, $x = 2.4 - 0.32y$
(c) $y = 3.24 - 0.517x$, $x = 0.314 - 1.03y$
(d) $y = 2.76 + 1.25x$, $x = 0.75 + 0.648y$
(e) $y = 1.29 + 0.431x$, $x = 0.265 + 0.275y$

8 Find the equations of the two regression lines and the product moment correlation coefficient for each set of data below.

(a)
x	2	3	4	5	6	7	8
y	2	4	5	7	10	12	16

(b)
x	6	8	9	12	16	17	19	21
y	23	22	19	17	15	12	8	5

(c)
x	3	7	8	13	16	18	22	27
y	14	19	21	29	33	36	40	45

(d)
x	4.5	6.2	9.4	10.3	14.1	16.2	19.5	23.7
y	6.4	17.3	7.0	9.1	21.4	13.8	5.5	8.2

9 The regression line of y on x, $y = a + bx$, passes through the points $(2, 3)$ and $(4, 4)$, and the line of x on y, $x = c + dy$, passes through the points $(2, 1.5)$ and $(3, 3)$. Find the values of a, b, c and d and so find the product moment correlation coefficient for these lines.

*10 (a) Use the summarised data given below to find the equations of the two regression lines for the original data.
   (b) Find (i) y, given that $x = 7.5$, (ii) x, given that $y = 4.2$.
   (c) Calculate the product moment correlation coefficient.

$$\sum x = 42.5, \quad \sum x^2 = 264.75, \quad \sum y = 66, \quad \sum y^2 = 844,$$
$$\sum xy = 369, \quad n = 7$$

11 A small A level class is given tests in mathematics and physics with the results shown in the table below.

Pupil	1	2	3	4	5	6	7	8
Mathematics (x)	48	81	65	68	52	56	67	75
Physics (y)	52	69	72	46	48	75	78	83

Calculate:

(a) Spearman's rank correlation coefficient,
(b) the product moment correlation coefficient for these two sets of marks.

12 Calculate Spearman's rank correlation coefficient (ρ) and the product moment correlation coefficient (r) for each set of data below.

(a)
x	30	70	50	60	25
y	40	50	60	80	70

(b)
x	5	8	3	1	4	8
y	2	1	1	5	6	3

(c)
x	12	10	9	7	5	2
y	4	8	13	19	23	28.

Intermediate

1 A pupil takes eight subjects in the summer examinations. The school examinations officer, curious about the relation between performance and the temperature at the time of writing, records the following figures for the pupil mentioned above.

Subject	1	2	3	4	5	6	7	8
Temperature, $x(°C)$	15.5	17	16	19	23	21	20	19.5
Mark, y	53	71	58	55	52	53	66	61

Calculate both Spearman's rank correlation coefficient and the product moment correlation coefficient and comment on their values. Suggest improvements to this experiment.

2 Find equations for the regression lines of y on x and x on y using the normal equations for each of the following sets of data. Plot the points and draw the lines for each set of figures.

(a)
x	1	2	3	4
y	3	7	9	13

(b)
x	2	4	6	8
y	12	8	7	4

(c)
x	1	4	5	8	9	13	16
y	3	9	8	15	18	24	27

(d)
x	5	10	15	20	25	30
y	10	9	5	5	1	2

3 You are given two sets of summarised data below. Use these to find, in each case, the equations of the regression lines of y on x and x on y. Do this in two ways:

  (i) using the normal equations,
  (ii) using the regression coefficients S_{xy}, S_x^2, S_y^2.

  (a) $n = 8$, $\sum x = 40$, $\sum y = 64$, $\sum x^2 = 260$, $\sum y^2 = 652$, $\sum xy = 410$
  (b) $n = 6$, $\sum x = 42$, $\sum y = 60$, $\sum x^2 = 424$, $\sum y^2 = 790$, $\sum xy = 270$

4 The equation of the regression line of y on x is $y = a + bx$. Calculate the equations of the following lines:

  (a) the slope of line is 0.3 and it passes through the point $(2, 1)$,
  (b) the line passes through the points $(3, -0.65)$, $(4, -1.15)$,
  (c) the line is parallel to the line $2y + 4x = 6$ and passes through the point $(2, -3.5)$,
  (d) the line has a gradient of $\frac{1}{5}$ and passes through the point $(1, 2.6)$,
  (e) the line passes through $(0, 0)$ and $(4, 5)$.

5 Find the equation of the line of regression of y on x from the set of readings shown.

x	2	3	4	5
y	1	4	6	10

6 (a) Plot the points shown below on a scatter diagram.

x	55	62	66	71	73	76	81	85	90	93
y	14	16	19	17	22	26	29	31	30	34

(b) Find the equations of the two regression lines and draw them on the diagram. Label the lines clearly.
(c) Use the lines you have drawn to find:
    (i) the value of y when $x = 70$,   (ii) the value of x when $y = 30$.

Compare these values with those calculated from the equations found in (b).

7 The variables x and y are related to each other as shown by the figures below.

x	2.8	2.9	3.2	3.4	3.6	3.9	4.2	4.6
y	5.6	5.3	5.2	4.9	4.6	4.5	4.4	3.8

(a) Plot the points on a scatter diagram.
(b) From the figures, find the equations of the two regression lines and plot them on the diagram.
(c) Find, from the lines, (i) y when $x = 4.5$, (ii) x when $y = 3$ and compare these values with those calculated from the regression lines.

8 Two employees, in their first week of a new job, record the times they take to complete the same task on each of the first five days. The times, in hours, are shown in the table below.

P_1	5	4.8	4.3	3.9	3.5
P_2	3.1	4.2	4.9	2.7	3.4

(a) Calculate both Spearman's rank correlation coefficient and the product moment correlation coefficient for these figures.
(b) What can you say about:
    (i) the correlation implied by the two coefficients,
    (ii) the skills and learning ability that P_1 and P_2 bring to the job?

9 The maximum speeds of service, in km/hour, achieved by a group of professional tennis players are compared with their heights in the table below.

Speed, S(km/h)	191	183	188	197	192	189	195
Height, H(m)	1.62	1.67	1.74	1.77	1.81	1.88	1.92

(a) Calculate (i) Spearman's rank correlation coefficient, (ii) the product moment correlation coefficient for these figures.
(b) Comment on the values produced in (a).
(c) Use the regression line of S on H to find an estimate of the maximum speed which a man of height 1.79 m might achieve.

10 The amount spent on advertising built-in kitchens and the annual profits (both in thousands of pounds) are carefully monitored by the manufacturers over a seven year period, producing the following results.

Advertising (£1000s) A	12	16	23	27	32	33	40
Profits (£1000s) P	437	435	436	439	443	449	464

(a) Plot a scatter diagram to illustrate these figures.
(b) Calculate Spearman's rank correlation coefficient.
(c) Draw the regression line of P on A.
(d) What figures for profits are obtained from the graph and the regression line equation for advertising costs of £38 000?

*11 From the data given below, calculate in each case:

(i) $\sum x$, $\sum x^2$, $\sum y$, $\sum y^2$, $\sum xy$, $\bar{x}$, $\bar{y}$,
(ii) S_{xy}, S_x, S_y,
*(iii) Spearman's rank correlation coefficient.

(a)

x	4	7	11	12	16	21	26	32
y	12	13	10	5	11	7	4	2

(b)

x	4.5	6.2	3.1	4.8	5.6	2.9	3.7	4.1
y	0.84	0.62	0.71	1.03	0.95	0.88	0.79	0.66

12 For the following data:

(i) find the equations of the regression lines of y on x and x on y,
(ii) calculate the values of y and x requested from the appropriate equations.

(a)

x	0.058	0.071	0.068	0.074	0.080	0.087	0.094
y	0.35	0.31	0.29	0.36	0.33	0.37	0.34

Calculate y when $x = 0.062$ and x when $y = 0.305$.

(b)

x	78.5	61.6	68.4	65.3	60.8	64.5	62.6
y	1.34	1.58	1.66	1.72	1.75	1.69	1.64

Calculate y when $x = 70$ and x when $y = 1.6$.

13 A class of ten A level pupils is set a test on Monday morning. The teacher feels the results are below expectations and spends time during the week revising the subject matter. A second test is given on Friday. One pupil missed the Friday test because of illness. All the marks are percentages, as shown in the table below.

Pupil	1	2	3	4	5	6	7	8	9	10
Monday, M	59	53	56	62	57	54	48	55	60	58
Friday, F	63	59	66	62	60	51	46	55	59	Absent

(a) Plot a scatter diagram for the first nine pupils.
(b) Calculate the equation of the regression line of F on M and so estimate a mark for the pupil who missed the Friday test.

14 The breaking strains, x newtons, of ten random samples of household string are shown below. The maker of the string then adds a small percentage of a new synthetic fibre to each type of string and tests the new products. The new breaking strains are recorded below as y newtons.

	1	2	3	4	5	6	7	8	9	10
x (N)	1955	2014	2235	2428	2599	2735	2844	2901	2935	3018
y (N)	2010	2108	2288	2527	2943	2881	3006	3155	3220	3342

Calculate:

(a) Spearman's rank correlation coefficient,
(b) the product moment correlation coefficient.

*15 A school advertises for a new English teacher. Eight applicants are invited to interview, which is carried out by the head teacher (H) and the head of English (E). Each applicant is given a mark out of 10 by both interviewers. The higher the mark, the better the applicant.

Applicant	1	2	3	4	5	6	7	8
H	4	6	5	7	6	3	2	6
E	3	3	6	5	4	1	3	7

Calculate Spearman's rank correlation coefficient and comment on its value.

Advanced

1 A new, specialist pumpkin fertiliser, Giantgro, is put on the market. The manufacturer used the following test before the product was launched. A measured quantity of liquid Giantgro (in ml to the nearest ml) was fed to ten pumpkin plants grown next to each other in conditions which were, as far as was possible, the same for each plant. All the pumpkins were harvested at the same time and their masses (in kg to the nearest kg) were recorded as shown.

Vol. of Giantgro, x (ml)	2.4	2.8	3.2	3.6	4.0	4.4	4.8	5.2	5.6	6.0
Final mass, y (kg)	1.8	2.1	2.1	2.6	2.9	3.3	3.8	4.3	4.6	5.0

(a) Plot a scatter diagram, find the point $M(\bar{x}, \bar{y})$ and draw a line of best fit through M.
(b) Find the equation of the regression line of y on x for these figures.
(c) From the equation, and from the line of best fit, find estimates for the final mass of a pumpkin which was fed 5.4 ml of Giantgro.
(d) What assumption would you be making if you tried to estimate from the line the final mass of a pumpkin which had received 8.5 ml of Giantgro?

2 Eight students taking A level mathematics took a trial examination in January and the public test in June. The two sets of marks are shown below.

Student	1	2	3	4	5	6	7	8
Trial mark, x	38	41	46	50	57	65	72	86
A level mark, y	42	40	49	49	61	67	75	84

(a) Draw a scatter diagram to illustrate these marks.
(b) Find $M(\bar{x}, \bar{y})$ and through it draw an estimated line of best fit.
(c) Find the equation of the regression line of y on x.
(d) Write down an estimated A level mark using (i) the line of best fit and (ii) the equation of y on x, for a trial mark of 44.
(e) One student missed the trial, but scored 62 in the A level examination. Explain briefly how you might find an estimate from the figures above for his trial mark.

*3 The points below relate the variables x and y.

x	1	A	$2A$	10
y	2	5	$2A$	7

A is a positive integer and $\sum xy = 123$.

(a) Calculate the value of A.
(b) Show that the equation of the regression line of y on x is $2y = x + 5$.

4 A personnel manager, interviewing job applicants, uses two test results to help her make decisions. The tests assess spatial abilities (S) and fluency in language (L). The scores for the tests are shown below for ten applicants. The spatial test is out of 30 and the language test is out of 50.

Applicant	1	2	3	4	5	6	7	8	9	10
S	7	3	12	16	8	10	15	12	17	15
L	31	26	38	33	29	34	40	32	44	39

(a) Calculate a value for the product moment correlation coefficient from these results.
(b) What does this value signify?
(c) Find the equations of the regression lines of L on S and S on L.
(d) If an applicant scored 14 marks on the spatial test, but missed the language test, estimate from the appropriate equation what his score might have been.

5 Thirty pairs of values of two variables S and T produce the following summary.

$$\sum S = 106, \quad \sum T = 151, \quad \sum S^2 = 478, \quad \sum T^2 = 831, \quad \sum ST = 560$$

(a) Calculate (i) the equation of the regression line of T on S, (ii) the product moment correlation coefficient.
(b) Given that the regression line of S on T is $S = c + dT$, use the figures you already have to calculate the value of d.

6 (a) Under what conditions would you be able to use Spearman's rank correlation coefficient but not the product moment correlation coefficient?
(b) Show that, if the variables X and Y have four pairs of values as below, then Spearman's coefficient is -1. A and B are positive integers.

X	A	$A+B$	$A+2B$	$A+4B$
Y	$A+10B$	$A+8B$	$A+6B$	$A+4B$

(c) If $A = 1$ and $B = 2$, calculate the value of the product moment correlation coefficient.

7 Five pairs of values of the variables x and y are shown in the table.

x	1	P	$P+1$	4	5
y	5	$Q-1$	4	3	Q

(a) You are given that $\sum x^2 = 55$ and $\bar{y} = 3$. Calculate the values of P and Q.
(b) Calculate (i) Spearman's rank correlation coefficient, (ii) the product moment correlation coefficient.

8 A television competition requires competitors to place six antique objects in order of value, the most expensive first. The objects' true values, in order, with the most expensive first, are O_1, O_2, O_3, O_4, O_5, O_6.

(a) The ranking attempt by one competitor was the following assessment, with most expensive first: O_2, O_4, O_1, O_3, O_6, O_5. Calculate Spearman's rank correlation coefficient for this attempt.
(b) Another competitor chose a ranking which gave this coefficient the value 0.6. Find any one possible order of the six objects needed to achieve this.

9 Eight athletes run a 100 m race and their times are recorded in the table below. They are then given an intensive two-week training course, after which the race is re-run, giving the second set of times shown.

Athlete	1	2	3	4	5	6	7	8
Time, x (s)	10.9	11.3	11.0	10.7	10.8	10.5	11.2	11.1
Time, y (s)	10.8	10.9	10.9	10.7	10.9	10.4	11.1	11.0

(a) Find the equation of the regression line of y on x.

(b) One athlete competed in the first race with time 10.9 seconds, but was injured for the second. Use the equation to estimate the time in which she might have run the second race.

(c) Three athletes missed the first race because of examinations. They took part in the training and ran in the second race, their times being 10.9, 11.2 and 11.3 seconds. Find the equation of the regression line of x on y and use it to estimate what their average time could have been in the first race.

10 Following an outbreak of the terrible white spot disease, several fruit farmers in the Coxshire district have been unable to sell their apples. The number of tonnes of fruit affected seems to depend on the rainfall in the previous winter months (November to February). The results for six farms are shown in the table below.

Rainfall, x (cm)	22.7	26.7	28.1	29.3	31.6	35.8
Mass affected, y (T)	21	36.3	32.5	31	49.5	51

(a) Draw a scatter diagram to illustrate these figures.

(b) Find the equation of the regression line of y on x, using the normal equations, with the following approximate figures:

$$\sum x = 174, \quad \sum y = 221, \quad \sum x^2 = 5156, \quad \sum xy = 6657$$

Draw this line on the scatter diagram.

(c) Using the line on the diagram, find an estimate for the tonnage affected if the rainfall had been 25.6 cm.

11 As part of a practical exercise in statistics, Miriam and Peter were shown photographs of 11 people and asked to estimate their ages. The actual ages and the estimates made by Miriam are shown below.

Actual age, x	86	55	28	69	45	7	17	11	37	2	78
Miriam's estimate, y	88	60	35	77	50	8	15	6	49	2	85

(a) Draw a scatter diagram of Miriam's estimate, y, and the actual age, x.

(b) Calculate the equation of the regression line of Miriam's estimate on actual age.

(c) Draw this regression line on your scatter diagram. Draw also the line $y = x$.

Comment on Miriam's estimates.

(d) The equation of the regression of Peter's estimate, z, on actual age is $z = 0.894 + 0.869x$. Draw this line on your scatter diagram and comment, as far as possible, on Peter's estimate, z.

(e) Calculate the product moment correlation coefficient between Miriam's estimate, y, and Peter's estimate, z.

Assume that $\sum z = 388$, $\sum yz = 24983$, $\sum z^2 = 20722$. Interpret your result.

(f) State, without further calculation, the value of the correlation coefficient if Miriam's estimate had been converted from years to months and Peter's estimate had been left in years.

(g) Comment on the appropriateness or otherwise of calculating a regression line in part (b) and a correlation coefficient in part (e).

[AEB]

Revision

1 For the set of figures given below, calculate:

(a) $\sum x$, $\sum y$, $\sum x^2$ $\sum y^2$, $\sum xy$, $\bar{x}$, $\bar{y}$,

(b) the equations of the lines of regression of y on x and x on y,

(c) the regression coefficients S_x^2, S_{xy}, and S_y^2.

x	1	2	3	4	5	6	7	8	9	10
y	19	18	16	15	13	11	12	10	9	9

2 (a) Using the data below, draw a scatter diagram.

x	2	5	8	11	14	17	20	23
y	1	2	5	7	6	10	10	11

(b) Find $M(\bar{x}, \bar{y})$ and use this to draw the two regression lines.

(c) Use the lines to find the value of (i) y when $x = 10$, (ii) x when $y = 8$.

3 You are given two sets of summarised data below. Use them to find the equations of the lines of regression of y on x and x on y. Do this using (i) the normal equations, (ii) regression coefficients.

(a) $\sum x = 64$, $\sum y = 128$, $n = 8$, $\sum x^2 = 680$, $\sum y^2 = 2110$, $\sum xy = 926$

(b) $\sum x = 65.4$, $\sum y = 141$, $n = 7$, $\sum x^2 = 803.18$, $\sum y^2 = 3293.66$, $\sum xy = 1024.79$

4 Plot a scatter diagram for the points shown below.

x	1.7	2.3	4.1	4.8	5.3	5.9	6.3	6.6	7.2	7.6
y	64	56	59	56	47	55	57	51	41	42

Calculate the equation of the regression line of x on y for these figures and use it to find the value of x when $y = 61$.

5 (a) For the points shown below, the line of regression of y on x passes through the point (4, 1). Use this to find the equation of this line.

x	2	4	6	8
y	1	3	7	9

(b) Given that the regression line of x on y passes through the point (1, 3), find its equation in the same way as you did above.

6 Two motoring journalists are chosen to comment on six new car models. Each assesses fifteen different qualities for each car and gives a score of one or zero for that quality. The table shows the scores awarded for the cars by the two journalists (J_1 and J_2).

Car	1	2	3	4	5	6
J_1	12	8	11	13	9	10
J_2	10	9	13	12	7	8

Calculate both Spearman's rank correlation coefficient and the product moment correlation coefficient from these figures and comment on your answers.

7 The data shown in the table below is compiled from the sales figures for two shops which are part of a nationwide department store chain. The two shops are of roughly the same size, situated in areas of similar population. They sell the same goods at identical prices.

Article	1	2	3	4	5	6	7	8
Number sold $\times 100$								
Shop 1	15	21	34	26	19	23	31	28
Shop 2	24	18	20	38	32	41	35	28

Calculate Spearman's rank correlation coefficient. Mention possible reasons for its very low value.

*8 From the data in the table below, calculate:

(a) the product moment correlation coefficient,
(b) the equations of the two regression lines.

x	15.1	17.5	18.2	14.7	14.8	16.4	16.3	17.1	16.8	15.9
y	2.9	1.4	1.2	3.3	2.7	3.1	1.8	2.4	3.4	1.6

9 A meteorologist, investigating various aspects of global warming, records the noon temperature on ten randomly chosen days, in the same month two years running, at the same seaside resort in eastern England. His figures are shown below.

Day	1	2	3	4	5	6	7	8	9	10
T_1 (°C)	18.5	19.6	18.8	20.1	20.5	23.1	22.8	23.4	22.7	22.4
T_2 (°C)	18.8	19.4	20.7	20.9	21.3	22.5	22.6	23.2	24.4	19.8

What is the value of Spearman's rank correlation coefficient for these figures? Find the equation of the regression line of T_2 on T_1 and hence estimate what, according to these figures, a temperature of $22\,^\circ$C in the first year would imply in the second year.

10 Eight people join a health and fitness club and are persuaded to go on a weight loss diet. Their masses, before and after a two-month period on the diet, are shown below.

Person	A	B	C	D	E	F	G	H
W_1 (kg)	101	98	89	98	93	100	113	94
W_2 (kg)	96	93	82	93	92	98	107	90

Calculate Spearman's rank correlation coefficient for these figures. Does its value support the statement 'everyone loses mass proportional to their original mass' at this club?

11 Six people apply to join a choir. They are auditioned by two judges (J_1 and J_2) and placed in a merit order by being assigned a letter A–F, where A is the best, F the worst, as recorded in the table.

Applicant	1	2	3	4	5	6
J_1	C	F	A	E	B	D
J_2	B	E	A	F	D	C

Calculate Spearman's rank correlation coefficient for these assessments.

12 Five pupils studying English and history score the marks shown in the table in an end-of-year examination.

Pupil	1	2	3	4	5
English (E)	55	64	58	68	70
History (H)	63	47	51	49	50

Calculate:

(a) Spearman's rank correlation coefficient,
(b) the product moment correlation coefficient for these two sets of marks.

13 For each set of summarised data below, calculate the product moment correlation coefficient, using the formula $r = \dfrac{S_{xy}}{\sqrt{(S_x^2 \cdot S_y^2)}}$.

(a) $n = 5$, $\quad \sum x = 15$, $\quad \sum y = 25$, $\quad \sum xy = 89$, $\quad \sum x^2 = 55$,
$\sum y^2 = 159$
(b) $n = 5$, $\quad \sum x = 40$, $\quad \sum y = 40$, $\quad \sum xy = 312$, $\quad \sum x^2 = 370$,
$\sum y^2 = 472$

(c) $n = 6,$ $\sum x = 105,$ $\sum y = 90,$ $\sum xy = 1400,$ $\sum x^2 = 2275,$
$\sum y^2 = 1420$

14 For each of the following sets of figures,

(i) calculate the covariance S_{xy} and the two variances S_x^2 and S_y^2,
(ii) find the equations of the regression lines of y on x and x on y, using these coefficients.

(a)

x	5	6	8	9	11	14
y	3	4	7	8	11	12

(b)

x	10	12	13	15	18	20
y	24	19	19	13	9	3

(c)

x	2	3	4	5	6	8	10	11	12	13
y	1.5	3	3	4.5	5	5.5	7	8.2	8.3	9

6

Sampling and estimation

6.1 Unbiased estimates, the *t*-distribution

Unbiased estimators

$$\hat{\mu} = \frac{\sum x}{n}, \quad (\hat{\sigma})^2 = \frac{1}{n-1}\left(\sum x^2 - \frac{(\sum x)^2}{n}\right)$$

$$= \frac{n}{n-1}\left(E(X^2) - \{E(X)\}^2\right)$$

Central Limit Theorem $\quad \bar{X} = N\left(\mu, \frac{\sigma^2}{n}\right)$

Basic

Abbreviations used:

> pdf probability density function
> crv continuous random variable

For help with revision, it may be useful to keep concise records of your answers to question 1.

1 (a) For a population with mean μ and variance σ^2, explain what you mean by the following terms:

(i) an unbiased estimator, (ii) a consistent estimator, (iii) the best, or most efficient, estimator, (iv) the standard error of the mean, (v) the sampling distribution of means.

(b) Explain briefly what you understand about the Central Limit Theorem.

(c) Write brief notes on each of the following sampling methods:

(i) simple random sampling, (ii) systematic sampling, (iii) stratified sampling, (iv) cluster sampling, (v) quota sampling.

(d) Explain what is meant by the following terms used in sampling:

(i) population, (ii) census, (iii) sample survey, (iv) target population, (v) sampling units, (vi) sampling frame, (vii) bias.

2 X and Y are random variables defined as $X = N(40, 64)$ and $Y = N(50, 36)$. Samples of size 8 and 4 are taken from X and Y respectively. Write down the following distributions.

(a) $\bar{X}$ (b) $\bar{Y}$ (c) $\bar{X} + \bar{Y}$ (d) $\bar{X} - \bar{Y}$
(e) $4\bar{X} + 2\bar{Y}$ (f) $5\bar{X} - 4\bar{Y}$ (g) $\bar{Y} - 2\bar{X}$
(h) $\frac{1}{5}(\bar{X} + \bar{Y})$ (i) $\frac{1}{2}\bar{X} + \frac{1}{4}\bar{Y}$ (j) $\bar{X} - 3\bar{Y}$

3 In each case below, a random sample of size n is taken from a normal population of known mean and variance. For each example, write down the distribution of the sample mean and calculate the required probabilities.

	Original population	Sample size	Calculate
(a)	$X = N(100, 16)$	10	(i) $p(\bar{X} < 99)$
			(ii) $p(98 < \bar{X} < 101)$
(b)	$X = N(50, 9)$	6	(i) $p(|\bar{X} - 50| < 1)$
			(ii) $p(\bar{X} > 51.5)$
(c)	$X = N(2, 0.09)$	25	(i) $p(\bar{X} < 2.015)$
			(ii) $p(\bar{X} > 1.85)$
(d)	$X = N(1000, 90\,000)$	100	(i) $p(\bar{X} < 943)$
			(ii) $p(955 < \bar{X} < 1045)$

4 For the samples shown below, calculate unbiased estimates of the population mean and variance. Give each set of answers to an appropriate degree of accuracy.

(a) 4.1   6.4   3.9   5.2   4.7   5.5   4.3   4.8   4.6   4.8   5.0   4.7
(b) 0.032   0.028   0.031   0.029   0.037   0.033   0.035   0.030   0.032   0.031

5 The masses of ten captured wild male pheasants are recorded below, in grams to the nearest 5 grams.

1045   985   1015   995   1225   1140   1080   1135   1215   1075

Calculate unbiased estimates of the mean and variance, to the nearest 5 g and the nearest whole number, respectively, of the male pheasant population from these figures.

*6 A mathematics teacher wishes to know estimates for the mean and variance of the marks of all students who take the GCSE examination for which her class is entered. The results for her class of 28 pupils are summarised as $\sum x = 1398$, $\sum x^2 = 70\,704$. What are her unbiased estimates for the population mean and variance, to three significant figures?

7 The heights, in millimetres, of a basketball squad are detailed below.

  1854   1930   1803   1880   2007   2057   2159   1956   1930   1905
  1981   2032

Calculate unbiased estimates for the mean and variance (to the nearest whole number) of the heights of the population of basketball players from these figures.

8 The figures below are a random sample from a population with mean μ and variance σ^2.

  A 3   3 $2A$ A 1   3   3 A A

You are given $\sum x^2 = 69$. Calculate unbiased estimates for μ and σ^2.

9 The readings below are random samples from populations with unknown mean and variance. For each set, calculate unbiased estimates of the population mean and variance.

  (a) 15   10   15   20   20
  (b) 4   4   3   4   3   6
  (c) 1.6   3   2.4   1.7   2.8
  (d) 5   7   4   7   8   5   6   4   5
  (e) 24   58   61   47   36   50   54   44   65   61
  (f) 4.6   7.1   3.9   6.6   6.2   5.7   4.8   5.0   5.4   5.2
  (g) 11   15   13   19   23   14   17   13   16   14
  (h) 3   2   7   4   5   8   3   6   5   4
  (i) 12   7   8   5   11   9   10   6   5   8
  (j) 4.7   4.1   5.3   4.4   3.9   5.6   6.2   4.8   4.5   4.0

10 From the data summaries shown below, calculate unbiased estimates of the population mean and variance.

  (a) $\sum x = 65,$ $\sum x^2 = 875,$ $n = 6$
  (b) $\sum x = 34,$ $\sum x^2 = 180,$ $n = 8$
  (c) $\sum x = 21.8,$ $\sum x^2 = 59.4646,$ $n = 8$
  (d) $\sum x = 44,$ $\sum (x - \bar{x})^2 = 26,$ $n = 8$
  (e) $\sum x = 215,$ $\sum (x - \bar{x})^2 = 42.5,$ $n = 10.$

*11 Small cartons of fruit juice are labelled as containing 200 ml of juice. A random sample of 12 cartons contained the following volumes (in ml to the nearest ml):

  203   205   200   198   201   203   199   202   204   201   203   198

Calculate unbiased estimates for the mean and variance of the volume in the total population of cartons.

12 The lengths of valve rods for a motor car engine are tested. The results (in cm to the nearest 0.01 cm) of a random sample of 20 rods were found to be:

31.30 31.25 31.31 31.28 31.26 31.30 31.29 31.28 31.27 31.26
31.31 31.27 31.26 31.29 31.30 31.28 31.31 31.28 31.26 31.30

Find unbiased estimates for the population mean and variance of the length of a rod.

13 The mass of an egg is used to determine whether it is sold as small, medium or large. A large egg is defined as one whose mass is normally distributed, with mean 78 g and standard deviation 5.4 g. A technician selects ten large eggs at random from stock and weighs each one. Write down unbiased estimates of the mean and standard deviation of the mass of this sample. What is the probability that its mass is greater than 80.5 g?

14 The continuous random variable X is distributed as $N(80, 16)$. A random sample of size four is taken from the population. What is the probability that the sample mean

(a) is less than 77,   (b) lies between 78 and 84?

15 In each part of this question, the results of testing a single sample of manufactured items is shown. Find an unbiased estimate for the proportion of faulty items in the population in each case.

Item	Sample size	Number of faulty items
(a) Safety razors	100	4
(b) Ballpoint pens	60	2
(c) $3\frac{1}{2}$ inch computer discs	50	2
(d) Metal can openers	80	2

16 For each set of data below, the random variable X is 'the number of "successes" in the sample'. Calculate the required probabilities using a normal approximation. Remember to use a continuity correction in each case.

	Sample size	p(success)	Calculate
(a)	100	0.3	(i) $p(X < 40)$
			(ii) $p(X > 19)$
			(iii) $p(24 < X < 33)$
(b)	200	0.2	(i) $p(X > 32)$
			(ii) $p(30 < X < 38)$
(c)	150	0.4	(i) $p(|X - 60| < 10)$
			(ii) $p(|X - 60| > 15)$

17 The masses, in grams to the nearest gram, of £1 coins are normally distributed as $M = N(35, 0.09)$. In a sample of 50 coins what is the probability that the mean mass is (a) less than 34.9 g, (b) less than 35.02 g?

18 In each part of this question, calculate the value of t. [d represents the number of degrees of freedom.]

(a)   (i) $p(T < t) = 0.975$, $d = 3$ (ii) $p(T < t) = 0.99$, $d = 10$
      (iii) $p(T < t) = 0.75$, $d = 2$ (iv) $p(T < t) = 0.95$, $d = 6$
(b)   (i) $p(T < 1.886) = 0.9$, $d = 2$ (ii) $p(T < 4.604) = 0.995$,
                                                          $d = 4$
      (iii) $p(T < 0.727) = 0.75$, $d = 5$ (iv) $p(T < 14.09) = 0.9975$
(c)   (i) $p(|T| > t) = 0.02$, $d = 4$ (ii) $p(|T| > t|) = 0.05$, $d = 8$
      (iii) $p(|T| > t) = 0.1$, $d = 5$
(d)   (i) $p(|T| < t) = 0.9$, $d = 3$ (ii) $p(|T| < t) = 0.95$, $d = 7$
      (iii) $p(|T| < t) = 0.8$, $d = 10$

Intermediate

1 In each example below, random samples of unknown size have been taken from a normal population with known mean and variance. Write down the distribution of the sample mean and find an estimate for the sample size in each case.

Population	Definition for $\bar{X}$
(a) N(100, 25)	$p(\bar{X} > 96) = 0.9633$
(b) N(2, 0.16)	$p(\bar{X} < 2.15) = 0.9332$
(c) N(15.6, 3.5)	$p(\bar{X} < 16.2) = 0.9242$
(d) N(10, 0.64)	$p(9.9 < \bar{X} < 10.1) = 0.4680$
(e) N(20, 4)	$p(|\bar{X} - \mu| < 0.2) = 0.2478$

*2 Random samples are taken from various distributions and the details of each are given below. Calculate the required probabilities, using a normal approximation. The mean of each sample is $\bar{X}$.

	Original distribution	Sample size	Calculate
*(a)	B(40, 0.65)	30	(i) $p(\bar{X} \geq 28)$
			(ii) $p(\bar{X} \leq 25)$
(b)	B(100, 0.4)	25	(i) $p(\bar{X} < 42)$
			(ii) $p(39 < \bar{X} < 42)$
(c)	B(14, 0.6)	4	(i) $p(\bar{X} < 9)$
			(ii) $p(|\bar{X} - \mu| < 1)$
(d)	Po(4)	30	$p(\bar{X} > 4)$
(e)	Po(16)	16	$p(\bar{X} < 18)$

3 A population consists of the integers 5, 10, 15. Construct a frequency distribution for the means of all possible samples of size two, taken without replacement. Verify that if μ, σ^2 are the mean and variance of the original population, then $E(\bar{X})$ and $Var(\bar{X})$, the mean and variance of the distribution of the sample means, are given by $E(\bar{X}) = \mu$,

$$Var(\bar{X}) = \frac{\sigma^2(N - n)}{n(N - 1)}, \text{ where } N \text{ is the population total, and } n \text{ is the sample}$$

size.

4   Three packets of flower seed, borage, nasturtium and cowslip, have average stated contents of 75, 45 and 145 seeds respectively. Thirty packets of each type were checked and gave the following results.

Borage $\sum x_B = 2270$, $\sum x_B^2 = 171\,864$
Nasturtium $\sum x_N = 1381$, $\sum x_N^2 = 63\,627$
Cowslip $\sum x_C = 4437$, $\sum x_C^2 = 656\,393$

(a) Find unbiased estimates of the mean and variance of each population.
(b) A second sample of 30 nasturtium seed packets was checked, the following results being obtained for the number of seeds per packet.

48 51 46 47 45 49 48 46 44 48
52 50 49 45 48 52 50 47 46 48
45 43 47 48 45 46 51 49 50 49

Calculate an unbiased estimate for the variance of this sample.

5   A tutorial college advertises that 80% of its A level students achieve a grade C or higher in the January examinations.

(a) From a sample of 30 of the college's students who take the January examination, what is the probability that the number achieving a grade C or higher is (i) more than 27, (ii) less than 20? Use a normal approximation to the binomial distribution.
(b) Calculate the same quantities using the distribution of the continuous random variable P_s, 'the proportion of "successes" in the sample'.

6   A sample of size 80 from a population is found to have a proportion of 'successes' of 0.44. The continuous random variable X is 'the number of successes in the sample'. Calculate:

(a) $p(X > 42)$
(b) $p(28 < X < 38)$
(c) the value of k if $p(X < k) = 0.8838$

*7   During the winter term in a school, the boys choose to play either rugby or soccer. The proportion of those choosing soccer is, on average, 70%. If S is the random variable 'the number of boys choosing to play soccer', use a normal approximation to find the probability that, in a random sample of 60 boys, more than 40 have chosen to play soccer. What is the probability that in this sample three times as many have chosen soccer rather than rugby?

8 A sample of 80 schoolgirls contained four whose hair was naturally red. R is the random variable, 'the proportion of redheads in the sample'.
   (a) Write down the distribution of R.
   (b) Calculate (i) $p(R \leq 0.055)$, (ii) $p(R \geq 0.085)$.
   (c) What is the probability that, in a second sample of 80 girls, five are redheads?

9 The two samples shown below are random selections from a population with unknown mean and variance.

$$\begin{array}{llllllllll}
\text{Sample 1:} & 4 & 2 & 2 & A & 4 & 4 & 2 & A \\
\text{Sample 2:} & B & 2 & 4 & 4 & 2 & 2 & 2 & 4 & 2 & (B-2)
\end{array}$$

You are given that A and B are positive integers, $\bar{x}_1 = 3.75$, $\sum x_2^2 = 102$, $s_2^2 = 1.2$. Find A and B and hence an unbiased estimate for the population variance from these two samples.

10 A candidate representing the Dinosaur political party sends out, at random, 150 questionnaires to people on the electoral roll in his constituency. One question asked if the person answering knew of the party.
   (a) If all the questionnaires were returned with this question answered, 24 stating 'yes', how many people on the roll of 12 450 could be estimated as knowing of the Dinosaurs?
   (b) If only 106 of the first set of questionnaires had been returned with 24 'yes' answers, find a second estimate for the number who knew of the Dinosaurs.

11 The sample shown below is from a normal distribution with unknown mean and variance, where A is a positive number. You are given that the sample mean is 4.

$$\begin{array}{llllllllll}
4.2 & 4.1 & 4.3 & A & 3.9 & 4.1 & A+0.3 & 3.9 & 4.0 & A+0.1
\end{array}$$

Find (a) A, (b) the sample variance, (c) an unbiased estimate for the population variance.

12 The Greenyard garden centre buys in rosemary plants and has found, over a long period, that 3% of the plants fail after being sold and have to be replaced. A sample of 50 plants is tested from a large batch which has been bought in. What is the probability that (a) more than 5% (b) less than 2% of the plants fail?

13 A shopkeeper receives a consignment of ten trays of lemons, each holding 36 lemons. In the first tray, two lemons are found to be damaged, and a second tray has three damaged lemons. Use these results to find two estimates of the proportion of damaged lemons in the whole consignment. Find the mean of these estimates. If the shopkeeper pays 8p per lemon and sells them at 18p each, selling only 72% of the undamaged fruits, what profit would he expect to make, to the nearest penny, from this batch?

14 A market research firm is asked by a company to report on the sales of one of its products in a certain town. Two samples of people were chosen at random and interviewed, the results being displayed below.

Sample 1: $n_1 = 200$, number buying the product $= 49$
Sample 2: $n_2 = 350$, number buying the product $= 110$

Taking the average of these results, what is the unbiased estimate of the proportion of people in the town who buy this product?

15 A normally distributed random variable X has a distribution with unknown mean and variance 36. A sample of size four from this distribution is tested and it is found that $p(\bar{X} < 124) = 0.9087$. Calculate the value of the population mean to three significant figures.

16 Calculate the following probabilities using the t-distribution tables.
(a) $p(T > -1.638)$, $d = 2$ (b) $p(T > -4.773)$, $d = 5$
(c) $p(T > -2.145)$, $d = 14$ (d) $p(T < -2.120)$, $d = 16$
(e) $p(T < -7.173)$, $d = 2$

17 Calculate the following probabilities using the t-distribution tables.
(a) $d = 5$: $p(2.015 < T < 4.773)$ (b) $d = 3$: $p(2.253 < T < 5.481)$
(c) $d = 7$: $p(-1.415 < T < 3.499)$ (d) $d = 20$: $p(-1.325 < T < 3.552)$
(e) $d = 15$: $p(-2.602 < T < 2.947)$

18 The speed, in km h^{-1}, at which a fast bowler bowls is recorded 12 randomly chosen times with the following summarised results:
$\sum x = 1027.1$, $\sum x^2 = 88\,035.33$. Calculate a 95% symmetric confidence interval for the mean speed of all his bowling.

Advanced

*__1__ A watch company manufactures small steel rods and bearings for the winding mechanisms of watches. The diameters of the rods are normally distributed with mean 0.44 mm and standard deviation 0.1 mm, and the internal diameters of the bearings are also normally distributed with mean 0.47 mm and standard deviation 0.12 mm.
(a) Calculate the probabilties of the following random selections:
    (i)   the diameter of a single rod exceeds 0.465 mm,
    (ii) the combined diameters of 50 rods are less than 20.5 mm,

(iii) the diameter of a single bearing is less than 0.445 mm,
(iv) a single rod will be too wide to fit into its bearing.

(b) Samples of 100 rods are tested each week. What is the distribution of the sample mean diameter? What is the probability that the mean diameter of a sample is between 0.42 and 0.45 mm?

2 (a) From the population {3, 6, 9, 12}, samples of size three are taken (i) with replacement (ii) without replacement. Write down frequency distributions for the means of all possible samples of size three in each case.

(b) If μ, σ^2 are the mean and variance of the original population, verify that the mean and variance of the frequency distribution in (a) (ii) are, respectively,

$$\mu, \quad \frac{\sigma^2(N-n)}{n(N-1)}, \quad \text{where } N = 4, n = 3$$

3 (a) A fair coin is tossed 200 times. What is the probability that (i) 115 heads, (ii) more than 105 heads are thrown? If the coin had been biased so that p(heads) = 0.4, what would have been the probability that (iii) less than 71 heads, (iv) between 85 and 90 heads, inclusive, were thrown?

(b) 5000 £1 coins are delivered to a branch of the Eastshires Bank. The mass of a coin, in grams to the nearest 0.01 gram, is normally distributed as N(38.00, 0.04). If a sample of 10 coins is taken from this batch, what would you expect to be the value of the coins, to the nearest whole number, whose mass was less than 37.95 g?

4 A continuous random variable X has a probability density function f(x), where

$$f(x) = \begin{cases} kx(1+x) & 0 < x < 2 \\ 0 & \text{otherwise} \end{cases}$$

(a) Find the value of k.
(b) Calculate the mean and variance of X.
(c) Find the probability that the mean, $\bar{X}$, of a random sample of 50 values of X will be less than 1.38.

5 A continuous random variable X has a probability density function f(x), where

$$f(x) = \begin{cases} \frac{1}{4}(1+2x) & 0 < x < 1 \\ \frac{3}{4} & 1 < x < \frac{5}{3} \\ 0 & \text{otherwise} \end{cases}$$

(a) Verify that f(x) is a probability density function.
(b) Find the mean and variance of X.
(c) Find the probability that a random sample of 50 observations of X will have a mean ($\bar{X}$) which is less than 1.

6 (a) Random samples of size 60 are taken from a normal population which is defined as $X = N(10, 64)$. $\bar{X}$ is the crv 'the mean of the sample'. Find $p(\bar{X} < 9.1)$.
(b) The samples shown below are taken from a normal population with unknown mean and variance.

Sample 1:   2 4 3 5 6 2 3 5
Sample 2:   4 2 2 3 5 2 4 2 2 3

Calculate unbiased estimates of the population mean and variance for each sample and find the mean of these.

7 In a school of 1000 pupils there are 450 boys and 550 girls. The number of left-handed pupils found in two random samples each of boys and girls is given below.

Sample 1: 40 boys, 8 left-handed   Sample 2: 50 boys, 4 left-handed
Sample 3: 50 girls, 5 left-handed   Sample 4: 80 girls, 10 left-handed

From each sample, calculate unbiased estimates of the proportions of left-handed pupils. Find the average proportions of left-handed boys and girls in the school and find the mean of these two proportions, which will give a final unbiased estimate of the proportion of left-handed pupils in the school.

8 (a) Write down all the possible samples of size two which can be taken from the population {2, 4, 6}, (i) with replacement, (ii) without replacement, and hence construct frequency distributions for the means of the samples.
(b) Calculate the mean, μ, and variance, σ^2, of the original population and verify that the mean and variance of the distribution in (a) (ii) are

$$\mu, \quad \frac{\sigma^2}{n}\left(\frac{N-n}{N-1}\right), \quad \text{respectively,} \quad \text{where } N = 3, n = 2$$

Revision

1 The frequency distributions given below are random samples from populations whose mean and variance are unknown. Calculate unbiased estimates for the population mean and variance in each case.

(a)

x	2	4	6	8	10	12
f	4	9	17	30	23	18

(b)

x	4.5	4.8	5.1	5.3	5.4	5.7	6.0	6.2
f	2	6	13	24	38	47	45	36

(c)

x	5	7.5	10	12.5	15	17.5	20	22.5	25	27.5	30
f	12	19	32	51	48	37	30	22	18	14	13

2 A continuous random variable X is distributed as N(35, 25). What is the probability that, in a random sample of size 10, the sample mean is

(a) greater than 37,   (b) more than 36.4 or less than 33.1?

3 A continuous random variable X is normally distributed as $N(\mu, \sigma^2)$. A sample of ten observations is shown.

$$47 \quad 52 \quad 51 \quad 50 \quad 48 \quad 49 \quad 51 \quad 52 \quad 50 \quad 50$$

(a) Calculate the sample mean.
(b) Calculate $p(X > 51.4)$ (i) using the sample mean and variance,
    (ii) using unbiased estimates for the population mean and variance.

4 A woodturner uses a lathe to turn decorative acorns from identical short lengths of wood. The time he takes, in seconds to the nearest second, is normally distributed with mean 185 seconds and variance 144 seconds. A sample of 10 acorns is chosen at random. Write down the distribution of the sample mean and calculate the probability that this time is (a) greater than 190 seconds, (b) between 175 and 180 seconds.

5 The woodpigeon, common in the British countryside and garden, has, when fully grown, an overall length (in centimetres, to the nearest 0.1 cm) distributed with a mean of 41.2 cm and variance 7.6 cm. A sample of 50 birds is netted and measured. Find the probability that the mean length of birds in the sample is

(a) less than 40.5 cm,   (b) between 41.4 and 41.7 cm.

6 The masses of a random sample of 20 bird eggs are recorded below. They are in grams to the nearest 0.01 g.

$$2.15 \quad 2.14 \quad 2.12 \quad 2.16 \quad 2.15 \quad 2.14 \quad 2.16 \quad 2.18 \quad 2.15 \quad 2.15$$
$$2.17 \quad 2.13 \quad 2.15 \quad 2.17 \quad 2.16 \quad 2.18 \quad 2.19 \quad 2.16 \quad 2.18 \quad 2.17$$

(a) Calculate the sample mean mass.
(b) Calculate the sample variance using (i) $\dfrac{\sum x^2}{n} - (\bar{x})^2$,

    (ii) $\dfrac{1}{n}\sum(x - \bar{x})^2$.

(c) Calculate an unbiased estimate of the population variance.

7 The continuous random variable X is normally distributed as $N(\mu, \sigma^2)$.
A random sample of 50 observations of X gives $\sum x = 279.6$,
$\sum x^2 = 1569.6$. Calculate:

(a) the sample mean and variance,
(b) an unbiased estimate for the population variance.

*8 Samples of given size from normally distributed populations of known
means are shown below. Calculate the value of the population variance
in each part. $\bar{X}$ is the sample mean.

	Population	n	Definition for $\bar{X}$
(a)	$N(100, \sigma^2)$	16	$p(\bar{X} < 103) = 0.9772$
(b)	$N(80, \sigma^2)$	9	$p(\bar{X} < 82.5) = 0.9332$
*(c)	$N(10, \sigma^2)$	10	$p(\bar{X} > 10.1) = 0.4164$
(d)	$N(8.5, \sigma^2)$	30	$p(\bar{X} < 8.35) = 0.3304$

9 Samples from normal distributions with known means provide
probabilities as shown. The sample mean is $\bar{X}$. Calculate the value of k in
each part.

	Distribution	Sample size	Probability definition
(a)	$N(150, 64)$	25	$p(\bar{X} < k) = 0.0062$
(b)	$N(2, 0.49)$	40	$p(\bar{X} < k) = 0.7645$
(c)	$N(1000, 62\,500)$	200	$p(985 < \bar{X} < k) = 0.673$

10 The continuous random variable X has a probability distribution
function $f(x)$, where

$$f(x) = \begin{cases} kx & 0 < x < 1 \\ k(2x - 1) & 1 < x < 2 \\ 0 & \text{otherwise} \end{cases}$$

(a) Find the value of k.
(b) Find $E[X]$ and $V[X]$.
(c) In a sample of n random observations of X, it is found that
$p(\bar{X} < 1.51) = 0.9565$, where $\bar{X}$ is the sample mean. Calculate the
value of n.

11 Washout, a new washing powder, is put on the market after an intensive
advertising campaign. A market research firm is commissioned to test its
popularity. Two random samples of people are interviewed as detailed
below.

Sample 1: $n_1 = 350$, number buying Washout $= 49$
Sample 2: $n_2 = 400$, number buying Washout $= 72$

(a) From the means of these results calculate an unbiased estimate of the proportion of people who buy Washout.

(b) It is estimated that the countrywide buying population of washing powders is 21 million people. If Washout is only sold in 1 kg packets, at £1.85 per packet, what income is implied by the proportion calculated above? Give your answer to the nearest £1000.

(c) For the second sample, the continuous random variable X is defined as 'the number of people in the sample who buy Washout'. Find the probability that in this sample, the number of people who buy Washout is (i) more than 80, (ii) less than 82.

12 Thirty trays, each holding 36 eggs, are delivered to a supermarket. A single tray is checked and found to contain two bad eggs. Find, to the nearest whole number, an estimate for the number of bad eggs in this delivery.

6.2   Confidence intervals

Basic

1 Each set of readings below is a random sample from a normal population whose variance is given. Calculate in each case (i) the sample mean, (ii) a 98% central confidence interval for the population mean.

(a) 5 7 6 3 9 11 8 7 6 8 $\sigma^2 = 2.25$
(b) 11.6   10.4   12.5   11.2   11.9   12.3   12.8   11.4 $\sigma^2 = 1.44$
(c) 0.4   0.6   0.3   0.5   0.6   0.8   0.7   0.4   0.6   0.9 $\sigma^2 = 0.04$

2 For each part of this question, calculate (i) a 99%, (ii) a 95% confidence interval for the population mean. Each variable is normally distributed and the population variance is given.

(a) The length of a plastic plant tie has a variance of 0.5 mm. A sample of ten ties gives a mean length of 15.2 cm.

(b) The variance of the lengths of sticks of medium spaghetti produced by the Pastagrande company is 0.15 cm. A sample of 50 sticks has an average length of 25.4 cm.

(c) Hockey balls produced by the Fieldhock company have a variance in mass of 0.48 g. The average mass of a sample of 100 balls is found to be 156.3 g.

(d) The Greenyard garden centre sells fence posts in two lengths, short and tall. The tall posts have a standard deviation in their length of 11.6 cm, and a sample of 40 posts gave a mean length of 2.41 m.

3 A population is known to have a variance of 36. Samples of varying size are taken and the sample mean is calculated. For each sample detailed, calculate the required confidence interval for the population mean.

	Sample size	Sample mean	Confidence interval
(a)	100	50	(i) 95%, (ii) 99%
(b)	225	90	(i) 95%, (ii) 98%
(c)	64	40	(i) 95%, (ii) 99%

*4 A random sample of 35 bags of horticultural sand sold in a garden centre is found to have a mean mass of 25.2 kg. If the masses of the bags are normally distributed with variance 0.4 kg, find (a) 90% and (b) 98% confidence intervals (CI) for their mean mass.

5   A garage which services motorcycles records the time taken to change a tyre on 100 machines. The mean of this variable is found to be 8 minutes 12 seconds, with a standard deviation of 1 minute 30 seconds. Find a 95% confidence interval for the mean time it takes to change a tyre at this garage.

6   A supermarket sells bags of granulated sugar. A random sample of 40 bags has a total mass of 80.8 kg, with variance 0.03 kg. Calculate a 95% confidence interval for the mean mass of these bags.

7   Seville oranges are imported into the UK in January and are largely used for making marmalade. The masses, in grams to the nearest $\frac{1}{10}$ gram, of a random sample of 50 oranges gave the following summary:

$$\sum w = 8707.9, \quad \sum w^2 = 1\,642\,379.1$$

where w is the random variable 'the mass of an orange'. Calculate an unbiased estimate for the variance of the mass of an orange and use it to find a central 99% confidence interval for the mean mass of Seville oranges.

8   From random samples of varying size, taken from several populations, the proportions of unacceptable members are shown in the table below. Calculate in each case a symmetric confidence interval, at the required levels, for the proportion of unacceptable members of the whole population.

	Sample size	p_s	Significance level
(a)	40	25%	(i) 95%, (ii) 99%
(b)	100	$\frac{1}{5}$	(i) 90%, (ii) 98%
(c)	50	15%	(i) 95%, (ii) 98%
(d)	60	20%	(i) 95%, (ii) 99%
(e)	100	14%	(i) 90%, (ii) 98%

9   A sample of size 50 is taken at random from a population, and five are found to be defective. Find a central 90% confidence interval for the

proportion of defective members of the whole population. If the total population were 10 000, what range of numbers of defective members does this imply?

10 The proportion of 'successful' items in samples of various sizes is shown below. Find the required confidence interval for the proportion of 'successes' in the whole population.

	Sample proportion	Sample size	Significance level
(a)	0.1	100	95%
(b)	0.6	140	90%
(c)	0.25	60	99%
(d)	0.85	50	98%
(e)	0.95	200	97%

11 For each set of data below, calculate both one-sided 95% confidence intervals for the population mean.

	Sample size	Sample mean	Population distribution	Population variance
(a)	20	30	normal	36
(b)	100	3	not known	9
(c)	300	4.4	normal	46
(d)	150	15.3	normal	78

12 During a boring statistics lesson, nine pupils in a class of 24 were seen to be eating toffees. Find from this a 99% one-sided confidence interval, of the form $(0, p)$, for the proportion of toffee-eaters in the whole population of statistics classes at this level in this school, assuming that they were all being taught the same (boring!) lesson.

13 In each part of this question, the population variance is not known. Use the t-distribution to calculate (i) a 95% and (ii) a 99% symmetric confidence interval for the population mean.

	Sample size	Sample mean	Sample variance
(a) The mass of a cricket ball	10	156 g	8.2
(b) The mass of a pair of spectacles	8	36.9 g	3.4
(c) The mass of an egg	6	76.7 g	9.8
(d) The volume of wine in a bottle	12	749 ml	11.2
(e) The volume of toothpaste in a tube	20	75 ml	3.85
(f) The mass of a packet of lentils	5	514 g	12.3

Intermediate

1 A survey of 2000 randomly chosen individuals revealed that 720 bought the *Daily Shout*, a national newspaper. Find a central 95% confidence interval for the proportion of the whole population that buy the *Shout*. From a buying population of 20 million, what are the largest and smallest daily circulations predicted by this result?

2 A sample of 100 private houses in a town showed that 43 contained more than one television set. Use this sample result to find a 95% confidence interval for the proportion of such houses in the town containing more than one television set. If there are 9480 private houses in the town which contain, legally, at least one television set, what upper and lower limits does the confidence interval give for the number of such houses containing at least one television set?

3 A sample of 200 torch bulbs is tested and the details of their useful life-spans, in hours to the nearest hour, are shown in the following grouped frequency distribution.

Hours	0–50	50–70	70–90	90–100	100–105
Frequency	15	33	75	52	25

(a) Estimate the sample mean and variance.
(b) Plot a frequency polygon of these results.
(c) Use the sample variance to find an unbiased estimate of the population variance.
(d) Find a 99% confidence interval for the population mean.

4 The masses of 49 soldiers, chosen at random from an Army barracks, in kg to the nearest kg, are shown below.

```
76  81  72  79  75  80  74  74  76  75
78  80  79  74  73  72  78  77  71  80
78  72  76  74  78  79  81  80  85  90
77  73  70  78  71  80  78  71  74  77
75  76  74  75  72  79  71  76  77
```

Calculate:

(a) the mean and variance of this sample
(b) an unbiased estimate for the population variance
(c) a 95% symmetric confidence interval for the mean mass of the whole population of the barracks.

5 The diameters of the leads for a clutch pencil, which are normally distributed, are tested by measuring a sample of 20 leads, and the results, to the nearest 0.01 mm, are given in the table below.

0.51   0.52   0.48   0.49   0.50   0.51   0.49   0.47   0.51   0.50
0.50   0.52   0.49   0.50   0.51   0.49   0.48   0.50   0.52   0.51

Calculate:

(a) the sample mean and variance,
(b) an unbiased estimate for the population variance,
(c) a symmetric 98% confidence interval for the mean diameter of such leads.

From the sample figures, find the probability that the mean diameter of leads lies between 0.47 mm and 0.49 mm.

6 A light engineering company manufactures 700 door handles each day for a certain type of car. To set up a guideline, the daily output is tested on a certain day, and five handles are rejected as being below standard. Find a 95% central confidence interval for the proportion of handles which would be rejected from the total output by the quality control team. From your answer, what are the greatest and least numbers of handles you would expect to be rejected in a five-day working week?

*7 On their first day at an initial training centre, the masses, in kg to the nearest kg, of 25 randomly chosen Army recruits were recorded. Taking W to be the random variable 'the mass of a recruit', the following results were calculated for this sample.

$$\sum w = 2050, \quad \sum w^2 = 168\,402$$

Calculate the sample mean $\bar{W}$ and variance s^2, and also an unbiased estimate for the population variance. Assume that the masses of recruits are normally distributed. For a random sample of this size, use your sample mean and variance to find the probability that $\bar{W}$ exceeds 90 kg. Calculate a 90% confidence interval for the population mean mass.

8 It is common to find shops selling goods at a discount during a recession. During such a period, a survey of 200 shoe shops found that 76 sold a particular brand of shoe at a discount.

(a) Calculate an unbiased estimate for the proportion of all shops selling this brand of shoe at a discount.
(b) Show that a 99% confidence interval for this proportion gives limits of ±9% for this proportion.

9 In each part of this question, an unbiased estimate of a population proportion is given for a random sample of size n. For a symmetric 95% confidence interval to be within the stated percentage of the estimate, find n, to the nearest whole number, in each case.

	Proportion	Percentage limit
(a)	0.44	$\pm 2\%$
(b)	0.28	$\pm 5\%$
(c)	0.36	$\pm 3\%$
(d)	0.66	$\pm 10\%$

10 The Electrostar electrical company makes batteries for automatic cameras. If X is the random variable, 'the number of hours a battery lasts', and a sample of tested batteries gives the summary data below, calculate:

(a) $\bar{X}$,
(b) the standard deviation of the sample,
(c) an unbiased estimate of the population variance,
(d) a central 90% confidence interval for the population mean

$$n = 26, \quad \sum x = 303.2, \quad \sum x^2 = 3536.88$$

11 The mass, in grams, of a new make of tennis ball is known to be normally distributed as N(58, 0.81). One specification for the ball is that the 95% confidence interval for the mass of a ball should have a width of less than 1. How large a sample of balls must be tested to achieve this?

12 In each part of this question, find the smallest sample size for the required width of specified confidence interval.

	Distribution	Width to be less than	Confidence interval
(a)	N(40, 1.69)	0.8	95%
(b)	N(85, 2.4)	1.2	90%
(c)	N(22, 1.2)	0.4	99%
(d)	N(148, 7.1)	3.2	98%
(e)	N(70, 4.8)	0.75	96%

13 A sample of size n from the population N(μ, 5.8) is used to calculate a 95% symmetric confidence interval for the population mean. Find the least value of n if the maximum value given by this interval is to be less than 5% above the sample mean of 36.

14 A consignment of hockey sticks delivered to a sports retailer is checked on arrival and a sample of 30 sticks is found to contain one defective stick. Find a one-sided 98% confidence interval for the proportion of defective sticks in the consignment in the form $(0, p)$.

15 A statistics teacher gives extra lessons twice a week to help students prepare for their A level examinations. Over a period of six weeks, he selects five lessons at random and finds the total attendance was 42. Assuming that the numbers attending follow a Poisson distribution, find a 95% one-sided confidence interval, in the form $(0, p)$, for the average attendance per lesson.

16 A random sample of 15 fence posts from a garden centre was found to have a mean length of 1.84 m. The population standard deviation for these posts is taken to be 1.7 cm. Use the t-distribution to find a symmetric 95% confidence interval for the mean length of such posts.

17 A manufacturer of fishing lines tests 10 spools of line and finds their breaking strains, measured in grams, to be as follows:

$$506 \quad 485 \quad 491 \quad 494 \quad 488 \quad 502 \quad 500 \quad 501 \quad 495 \quad 496$$

Use the t-distribution to find a 90% symmetric confidence interval for the mean breaking strain of this line.

Advanced

1 A sample of 160 television remote-control switches is tested in the factory and 8 are found to be defective.
(a) Find an unbiased estimate for the proportion of defective switches in the total population.
(b) A second sample of 75 switches showed 3 to be faulty. Pool the results from these two samples to find another unbiased estimate for this population proportion.
(c) From your pooled estimate, calculate a 90% confidence interval for the proportion of defective switches in the whole population.
(d) A third sample of 80 switches showed 3 to be defective. If the 90% confidence interval for this sample were required to have limits within 2% of the estimated proportion, how would the number of switches tested have to be adjusted?

2 To estimate the number of hedgehogs in a certain area, a sample of 100 are trapped, marked and released. Three months later, from the same area, a second sample of 100 hedgehogs showed 32 to be marked. What estimate for the total number of hedgehogs in the area does this give? Mention any assumptions you might make in arriving at this figure.
At a 99% confidence level, what are the maximum and minimum numbers of hedgehogs in the area implied by these results, and what does this range beome at a 90% level?

3 The figures below show the numbers of pupils in five of the year groups in a school.

1st year	2nd year	3rd year	4th year	5th year
93	106	101	96	104

On a day near the end of a term, a local police officer visits the school to give a lecture. A random sample of 50 pupils at the lecture gives the following result:

1st year	2nd year	3rd year	4th year	5th year
8	12	7	10	13

(a) Find a central 90% confidence interval for the proportion of 4th year pupils at the lecture.
(b) Find a one-sided 90% confidence interval for the proportion of 4th year pupils at the lecture.
(c) What is the least number of 4th year pupils at the lecture implied by the answers to (a) and (b)?
(d) After these results were calculated, it was recalled that, on the day of the lecture, there was a measles epidemic. The school registers gave the following figures of pupils in attendance that day:

1st year	2nd year	3rd year	4th year	5th year
71	79	82	69	91

What would be your answers to (a), (b) and (c) using these figures?

4   A sample of size n is taken from a normally distributed population. The sample mean is 3.5 and the variance is $\frac{7}{4}$. From the sample variance, the unbiased estimate of the population variance is 2. Calculate

(a) the value of n, (b) $\sum x$,
(c) $\sum (x - \bar{x})^2$, (d) a 95% confidence interval for the population mean.

5   A generous shopkeeper sells sweets in bags supposed to hold 250 g, but usually gives his younger customers a few extra. The masses of sweets, in grams to the nearest gram, in 30 bags sold are displayed below.

258	259	262	260	257	255	259	256	254	258
262	259	261	258	263	259	263	256	261	258
260	259	257	259	261	263	258	262	261	260

(a) Calculate the sample mean and standard deviation of these figures.
(b) Find an unbiased estimate of the population variance.
(c) Write down a 98% symmetric confidence interval of the mean mass of the sweets he sells in this way.

6   The time, x seconds, taken for a jumbo jet aircraft to take off from rest is measured on several occasions, giving the results below.

Average time $= 24.666$ seconds, $\sum x = 1233.3$, $\sum x^2 = 30755.05$

Calculate:

(a) how many readings of the take-off time were recorded,

(b) the sample variance,
(c) an unbiased estimate for the variance of the take-off time,
(d) a 96% symmetric confidence interval for the average take-off time.

7   A market research company carries out an opinion poll before a local
election. Three candidates, A, B and C, are contesting the election, and
3000 potential voters were asked to name their preferred candidate. The
result was:

$$\text{A } 1200, \quad \text{B } 1400, \quad \text{C } 400$$

(a) What proportion of the people from this sample intend to vote for
candidate C?
(b) Using this result, estimate the proportion of people in the whole
population of voters who will vote for candidate C.

*8   Twenty people use a metal measuring tape to measure the width of a
small building as part of a statistics experiment. Their results (m) were:

14.25   14.27   14.28   14.23   14.24   14.25   14.26   14.27   14.25   14.24
14.24   14.26   14.25   14.27   14.24   14.25   14.27   14.26   14.26   14.24

To the nearest centimetre, give a 95% symmetric confidence interval for
the width of the building.

9   A catering company asked 50 randomly selected college students to state
the amount of money, x, which they spent daily on lunch, and the
results were summarised by $\sum x = 56.50$ and $\sum x^2 = 66.80$.
Calculate unbiased estimates of the mean and variance of the amount
spent daily on lunch by students at the college, giving your answers
correct to three significant figures.
Hence find a 90% confidence interval for the mean amount spent daily
on lunch, giving the end-points correct to the nearst $0.01.
Justify the use of the normal distribution in constructing the confidence
interval.
[UCLES]

Revision

1   From the figures given below, find in each case a symmetric confidence
interval for the population mean at the required significance level.

	Sample mean	Sample size	Population variance	Significance level
(a)	5.2	60	0.4	90%
(b)	48	150	73	95%
(c)	8.6	80	2.75	98%
(d)	126	100	225	99%

2   Each sample considered below is from a normally distributed population
    which has an unknown variance. Calculate in each case a 95% central
    confidence interval for the population mean.

    (a) 4   8   7   5   3   6   7
    (b) 10 9   14 11 12 15 13 12
    (c) 21 18 24 26 22 20 25 24 27 23
    (d) 7.6 8.2 5.9 5.1 6.6 6.1 6.4 5.8 6.8 5.7

3   The summaries shown below are from normally distributed populations
    with unknown means and variances. Find an unbiased estimate of the
    population variance and hence a confidence interval at the required
    significance level for the population mean.

	Sample size	$\sum x$	$\sum x^2$	Significance level
(a)	20	192	1938	99%
(b)	45	192.2	1729.96	95%
(c)	30	3696	455 604	98%

4   Samples of given size, from normally distributed populations, are
    described by the data below. For each set, calculate (i) the sample mean
    and variance, (ii) an unbiased estimate of the population variance, (iii) a
    90% symmetric confidence interval for the population mean.

    (a) $n = 15$, $\sum x = 57$, $\sum(x - \bar{x})^2 = 26.4$
    (b) $n = 10$, $\sum x = 75$, $\sum(x - \bar{x})^2 = 30.5$
    (c) $n = 9$, $\sum x = 40.3$, $\sum x^2 = 183.37$

5   A box of 250 vinyl ring reinforcements is checked and eight are found to
    be defective. A second packet of 500 rings has fourteen which are
    defective. Pool these results to find an unbiased estimate of the
    proportion of defective rings in the whole population.

6   In a school with over 1000 pupils, a teacher notices that four of her A
    level mechanics class of 16 are left-handed. Give a 98% one-sided
    confidence interval of the form $(0, p)$ for the proportion of left-handed
    pupils in the school.

7   A research worker checks the time taken in a factory to install an
    electronic component in a tape-recorder. The average time taken by 50
    randomly chosen installers is 48 seconds. From previous research, it is
    known that the standard deviation for this time is 3.4 seconds.
    Find 90% and 95% confidence intervals for the mean time taken to
    install this component throughout the factory.

8 The sample shown comes from a population with unknown mean and variance.

$$3.5 \quad 4.6 \quad 3.8 \quad 3.9 \quad 4.1 \quad 4.0 \quad 4.4 \quad 3.6 \quad 4.0 \quad 4.2$$
$$4.1 \quad 4.3 \quad 3.8 \quad 3.7 \quad 3.9 \quad 4.0 \quad 4.1 \quad 4.2 \quad 3.9 \quad 3.7$$

Calculate:

(a) unbiased estimates for the population mean and variance,
(b) a 95% central confidence interval for the population mean, using the answers obtained in (a).

9 At the start of an academic year, 110 new pupils join school A, while school B records 96 new pupils. The heads of languages in the schools discover that only 15 and 12 pupils respectively have studied Latin! Pool these results to find an estimate for the proportion of new pupils in the whole population who have studied Latin. The first school, from its own record, wishes to calculate a 95% confidence interval for the proportion of new pupils in the population who have learnt Latin.
What figures does it produce?

10 At a university, 32 new students enrol for a sports science degree. Their masses, in kg to the nearest kg, are recorded on arrival:

78 86 82 88 92 76 84 81 82 83 79 82 80 78 79 85
84 84 77 82 90 78 86 81 78 91 98 85 83 80 76 82

Calculate:

(a) the mean mass of this sample,
(b) an unbiased estimate for the variance of the mass of all the students entering university for this course,
(c) a 95% central confidence interval for this population mean mass.

11 Various sample proportions are given below. In each case calculate a 90% symmetric confidence interval for the population proportion.

(a) $n_1 = 150$, $p_s = 0.25$ (b) $n_2 = 120$, $p_s = 0.3$
(c) $n_3 = 60$, $p_s = 0.65$ (d) $n_4 = 100$, $p_s = 0.5$

*12 The masses, in grams, to the nearest gram, of 10 statistics textbooks are recorded below.

841 935 1151 1020 397 895 760 928 1008 856

Use these values to find a 95% central confidence interval for the population mean. When 120 assorted mathematics textbooks were weighed, 36 of them were found to have masses above 1 kg. What estimate does this give, as a 95% central confidence interval, for the population proportion of mathematics textbooks with masses less than 1 kg?

7
Hypothesis testing

7.1   Types of test

Basic

Questions 1, 2, 3, 4 and 5 are for simple practice. Use the test statistic Z and not the t-distribution to work through them.

1 In each question below, a single sample value, x, has been taken from a normally distributed population whose variance or standard deviation is given. Test each hypothesis at (i) the 1% level, (ii) the 5% level, (iii) the 10% level.

   (a) $x = 76$; H_0: $\mu = 72$, H_1: $\mu > 72$, $\sigma^2 = 9$
   (b) $x = 18$; H_0: $\mu = 22$, H_1: $\mu < 22$, $\sigma^2 = 4$
   (c) $x = 69$; H_0: $\mu = 76$, H_1: $\mu \neq 76$, $\sigma = 4.9$
   (d) $x = 44$; H_0: $\mu = 42$, H_1: $\mu \neq 42$, $\sigma^2 = 0.6561$
   (e) $x = 0.34$; H_0: $\mu = 0.28$, H_1: $\mu \neq 0.28$, $\sigma = 0.0343$
   (f) $x = 185$; H_0: $\mu = 170$, H_1: $\mu > 170$, $\sigma = 8.6$
   (g) $x = 308$; H_0: $\mu = 300$, H_1: $\mu \neq 300$, $\sigma^2 = 25$
   (h) $x = 2.75$; H_0: $\mu = 3.10$, H_1: $\mu < 3.10$, $\sigma = 0.135$
   (i) $x = 1250$; H_0: $\mu = 1190$, H_1: $\mu \neq 1190$, $\sigma = 29$
   (j) $x = 0.64$; H_0: $\mu = 0.7$, H_1: $\mu < 0.7$, $\sigma^2 = 0.001\,156$

2 One pupil scores 72% in an examination. The teacher now claims that, as the variance of all the results is 18, the average mark for this examination has risen above the average of 64% for the past five years. Test this assertion at the 5% level of significance.

3 The average mass of an apple from an orchard in Kent is thought to be 280 g, with a variance of 240 g^2. A single apple is picked at random and its mass is found to be 306 g. Does this indicate, at the 5% level of significance, that the average is more than 280 g?

*4 In a class of 24 students, the average mass, M kg, is thought to be normally distributed as $M = \text{N}(85, 7.4^2)$. One student, chosen at random, is found to have a mass of 66 kg. Does this, at the 5% significance level, show that the average mass is less than 85 kg?

5 The average diameter of a batch of ball bearings is thought to be 0.22 cm with a standard deviation of 0.00175 cm. When a single bearing is found to have a diameter of 0.224 cm, does this, at a 1% significance level, mean that the average diameter is not 0.22 cm?

6 A die is rolled 60 times and 12 sixes are recorded. Test, at the 10% level, using a normal approximation, to see if the die is biased in favour of sixes. Remember to use a continuity correction.

7 Over a period of several months, a shopkeeper calculates that she sells, on average, 20 packets of toffee per day, with a standard deviation of 2. On one day, after making this calculation, 16 packets are sold. At the 5% level of significance, does this justify the shopkeeper saying that the average has dropped below 20 packets per day?

8 In the following questions, a sample of the size shown is taken from a normal population of given mean and standard deviation. Test each pair of hypotheses at (i) the 5% level and (ii) the 10% level of significance.

	n	x	σ	H_0	H_1
(a)	4	25.8	2.9	$\mu = 28.6$	$\mu \neq 28.6$
(b)	49	7.32	2	$\mu = 6.88$	$\mu > 6.88$
(c)	9	242	16.5	$\mu = 250$	$\mu < 250$
(d)	80	4.05	0.78	$\mu = 3.92$	$\mu \neq 3.92$
(e)	6	65	2.3	$\mu = 63$	$\mu \neq 63$

9 A bar of chocolate is labelled as having a nominal mass of 100 g. A sample of 80 bars is found to have a mean mass of 100.1 g, and the standard deviation is 0.4 g. Calculate an unbiased estimate for the standard deviation of the whole population of these bars, and use it to find, at the 2% level of significance, whether the mean mass is more than 100 g.

10 A 'large' egg is defined as having, on average, a mass of 76 g with variance 7.3 g. A random selection of 12 eggs was found to have a mean mass of 73.8 g. Test, at the 1% level of significance, whether the population mean is less than 76 g.

11 Ink cartridges for Smoothrite ballpoint pens are claimed to be capable of producing a continuous line of writing of length 1.84 km, with standard deviation 195 m. When a sample of ten cartridges is tested, the average length of line is found to be 1.92 km. Does this indicate, at the 5% significance level, that the average line length claim is too low?

*12 The average mass of a male goldcrest, the smallest British bird, is 6.2 g. After a severe winter, a random sample of 32 male goldcrests is trapped and the mean and standard deviation of their masses are found to be

5.85 g and 0.93 g respectively. Do these figures show that the average mass of the population has fallen below 6.2 g? Test at the 2% level.

Intermediate

1 The mean length of rods produced for a car component is 22.05 cm, to the nearest 0.01 cm, with variance 0.0049 cm^2. A sample of 25 rods gives a mean length of 22.02 cm. Does this show, at the 10% level of significance, that the mean has changed?

*2 The volume of wine in a bottle at the Imbibest winery has a mean of 750 ml and a variance of 1.21 ml. A sample of 50 bottles has a mean of 749.7 ml. Does this signify a change in the population mean, at the 5% significance level?

3 The masses of 'large' chickens on a supermarket shelf are normally distributed with mean 1.8 kg and variance 8 g^2. When a batch of 10 chickens, chosen at random from the same shelf, is tested, the mean mass is found to be 1.802 kg. Show that, at the 5% significance level, the mean is greater than 1.8 kg.

4 Bird nuts are sold in bags labelled as containing 4 kg of nuts. A shopkeeper who sells these bags receives complaints from customers that the mass of a bag is often below 4 kg. He tests a random sample of 50 bags and finds the mean mass is 3.8 kg with variance 0.66 kg. He tests these results at the 5% significance level. What does he find?

5 In each part of this question you are given the variance of a normal population, together with data from random samples and hypotheses about the population mean. Test each sample, using the t-distribution where appropriate, to determine which hypothesis is to be accepted.

Pop.	Pop. variance	Sample size	Sample mean	H_0	H_1	Significance level
(a) P_1	4.8	27	5.57	$\mu = 5$	$\mu > 5$	5%
(b) P_2	5.6	20	6.1	$\mu = 5.7$	$\mu \neq 5.7$	5%
(c) P_3	26	27	63	$\mu = 61.1$	$\mu \neq 61.1$	10%
(d) P_4	13.8	100	17.9	$\mu = 18.6$	$\mu < 18.6$	10%

6 A car battery is guaranteed for 24 months, and the population variance of this life is known to be 4.3 months. Test the following hypotheses at the 1% significance level to see whether this guarantee should be reduced, if a random sample of 80 batteries has a mean life of 23.4 months.

$$H_0: \mu = 24, \quad H_1: \mu < 24$$

7 The Trurite pencil company produces 0.3 mm leads for one of their clutch pencils. The lengths, in mm, of two random samples of ten leads each are shown below.

S_1:   60.6   60.8   60.8   60.6   60.7   60.6   60.8   60.7   60.5   60.7
S_2:   60.8   60.7   60.9   60.6   60.8   60.9   60.7   60.9   61.0   60.7

Assuming a population variance of 0.0998, test the population mean twice, using the mean of each sample, at the 10% significance level, according to the following two hypotheses.

$$H_0: \mu = 60.88, \quad H_1: \mu < 60.88$$

*8 The spinner disc shown is divided into four (equal) quadrants. The arrow, when spun, is supposed to stop at random round the circle. When it is spun 200 times, it is found to stop above area D 65 times. Test this result, at the 2% significance level, to see if the arrow is biased towards stopping over area D.

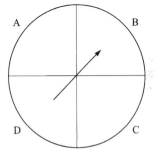

9 An automatic printing press can be set to run at various speeds. When the setting is 500 copies per hour, the rate is found to be normally distributed as $R = N(500, 123.8)$. At a 5% significance level, has the average speed increased if a test of 50 runs at a setting of 500 copies per hour produced a mean rate of 503 copies per hour?

10 The Pastagrande company produces standard spaghetti sticks where the length of a stick, L cm, is a normal variable such that $L = N(25.69, 0.03065)$. A suspicious, spaghetti-eating statistician suspects the sticks are getting shorter (a splendid alliteration!). He measures a sample of 20 randomly chosen sticks from several packets and finds the average length to be 25.6 cm. Does this show, at the 1% level of significance, that the sticks are getting shorter?

11 A racing pigeon owner keeps 158 birds and decides to enter them for an important race. They are given many practice flights over the course, and the average time, T hours, turns out to be a normal variable such that $T = N(2.76, 0.16)$. A friend of the owner recommends that he use a new type of birdseed, Speedifly, to improve the race times. After two months of using Speedifly, the average flight time for 40 birds, chosen at random from his coops, is 2.63 hours. If this value is tested at the 1% significance level, does it indicate a drop in the original average time of 2.76 hours?

12 A fan-assisted electric oven cooks a pie in T hours, which is a normal variable distributed as $T = N(0.74, 0.0625)$. The manufacturers improve the oven and claim that cooking times have been reduced. When a newly insulated oven is tested by cooking 45 pies similar to the one mentioned above, the average cooking time is found to be 0.69 hours. Show that this indicates a lowering of the cooking time for one of these pies at the 10% significance level, but not at the 5% significance level.

13 Two schools run a postal athletics competition. Ten randomly selected athletes from each school compete in various races which are timed by their coaches and the results are exchanged. For the 20 selected girls who ran the 100 m race, the two sets of times were as follows, in seconds to the nearest 0.1 second.

School 1:   12.4   12.4   12.3 T 12.5   12.4   12.4   12.4 T 12.4
School 2:   12.6   12.6   12.5   12.4   12.7   12.6   12.5   12.5   12.7   12.5

The lowest mean time of these 10 runners wins this event for their school. Calculate the value of T if the average time for S_1 is to be 0.2 seconds less than the average time for S_2.
The variance of the times for all the runners who were tested at S_2 is 0.02. Test the hypotheses below about the mean time for all tested runners at S_2 at the 1% significance level.

$$H_0: \mu_{s2} = 12.61, \quad H_1: \mu_{s2} < 12.61$$

14 The average score achieved by a rugby team over a season is 18 points per match with a variance of 4 points, and the scores are assumed to be normally distributed. In the first 10 matches of the new season, the average score was 19 points. Assuming the same variance, does this imply, at the 5% significance level, that the average score has gone up?

15 A certain model of car is thought to have a petrol consumption of 15 km/l. A random sample of 50 cars was tested, and their mean consumption was recorded as 14.6 km/l, with a standard deviation of 2.4 km/l. Does this mean, at the 5% significance level, that the mean population consumption is less than 15 km/l?

Advanced

1 A trout farm sells fish claimed to have an average mass of 500 g. To test this, 30 randomly chosen fish are weighed and their masses, in grams to the nearest gram, are recorded as shown.

513	504	490	497	509	510	508	502	504	489
505	513	501	506	496	505	500	498	491	503
488	504	502	500	495	499	504	508	501	506

Calculate the mean and variance of these masses and so find an unbiased estimate of the population variance. At the 5% significance level, do these figures imply that the population mean is more than 500 g?

2   The mass of a blackbird in Britain is a normal variable with standard deviation 4.2 g. Two random samples, S_1 and S_2, of 30 and 38 birds respectively are taken in different areas. In both of these areas, it is known that the average mass of a blackbird is 89.8 g.

(a) What is the probability that the mean mass of S_1 is less than 88 g?
(b) The mean masses of S_1 and S_2 are calculated to be 91.3 and 88.9 g respectively. Use the fact that the population mean is the same for both samples to test whether the mean mass of S_2 is significantly less than the mean mass of S_1, testing at the 2% level.

3   The members of the sixth form in a large school are about to elect one of three candidates, A, B and C, as their head prefect. From a random sample of 140 sixth formers, 42 say they will vote for candidate B.

(a) Find a 95% confidence interval for the proportion of sixth formers who will support candidate B.
(b) A member of staff (no doubt a cynical statistician!), thinks that the real proportion of voters who will support this candidate is 0.25. Test this hypothesis, at the 10% significance level, against the alternative hypothesis that the proportion is more than 0.25.

*4 A croupier at a casino is accused of using a die which is biased towards showing a six. The management tests the die by rolling it 500 times, and 98 sixes are recorded. At the 5% significance level, does this result support the complaint? To try to eliminate complaints, a new die is suggested. This die is also rolled 500 times, and 90 sixes are recorded. Calculate a 99% confidence interval for the proportion of sixes the die will show.
At the 5% significance level, does this die show bias towards a six?

5   In a particular summer, the proportion of left-handed students at a university is known to be 14%. During one of the final mathematics examinations, a bored invigilator walks round the room, counting the left-handed students, finding there are 16 out of 120. Does this mean, he speculates, that the proportion of left-handers is in fact less than 14%? He takes out his pocket calculator as soon as the examination is over (he is, after all, a statistician) and tests this at the 5% level.
What does he find?
Later on, he wonders what would have been the case if the proportion of left-handers in the university had been 13%. What is his second answer, if his alternative hypothesis is 'the proportion is not equal to 13%'?

6 A task on a car assembly line takes an average of T minutes, where T is a normal variable defined as $T = N(8.4, \sigma^2)$. The management arrange a work-study course for 50 randomly chosen members of their assembly staff and then time each one on this task. The result is an average time of 7.9 minutes, with a variance of 2.95. Does this represent a reduction in the original mean of 8.4 minutes, if it is tested at the 2% significance level? What is the maximum mean value, to three significant figures, of the average of the 50 staff from the course, if a reduction in the original mean is to be significant at the 5% level?

7 A tetrahedral die with faces marked 1, 2, 3 and 4 is rolled 200 times and the scores recorded as follows:

Score	1	2	3	4
Frequency	44	68	41	47

The following hypotheses are put forward:
H_0: $p(2) = 0.25$, H_1: $p(2) > 0.25$. Test, at the 1% significance level, to see if H_1 is justified.

8 A small estate car with a 1.6 litre engine has an advertised fuel consumption of 13.629 km/l, which is normally distributed with variance 1.96. When the engine is modified to a 16-valve system, the performance of the car goes up, but the fuel consumption is stated to rise by 2%, the variance remaining constant. A random sample of 40 16-valve cars is tested, giving an average consumption of 13.098 km/l. Does this imply that, at a 5% significance level, the consumption of the 16-valve engine is more than 2% above that of the original 1.6 litre engine? If a single 1.6 litre car were to produce a consumption of 14.868 km/l, would this imply that the original claim was understated, at a 5% significance level?

9 The volumes, in millilitres, of soft drink in a random sample of 30 bottles from a large batch were as follows:

```
499  525  498  503  501  497  493  496  500  495
502  499  496  492  497  492  503  505  501  498
493  496  502  495  496  497  501  496  499  513
```

(a) Stating clearly your null and alternative hypotheses, test the claim that the mean volume of soft drink in bottles in the batch is at least 500 ml. Use a 5% significance level. Assume that, for the batch, volumes are normally distributed with a standard deviation of 3 ml.
(b) Is there any reason to doubt the assumption that, for the batch, volumes have a standard deviation of 3 ml? Explain your answer.

(c) State, giving a reason, whether your conclusion in part (a) would be affected if

(i) the distribution of the volumes in the batch was not normal,

(ii) the standard deviation of bottles in the batch was greater than 3 ml.

[AEB]

Revision

1 Test each of the following pairs of hypotheses, which are based on a single sample value from a population with known variance.

	Sample value	Hypotheses	Variance	Significance level
(a)	69	$H_0: \mu = 65$ $H_1: \mu > 65$	12	5%
(b)	246	$H_0: \mu = 260$ $H_1: \mu \neq 260$	38	5%
(c)	0.471	$H_0: \mu = 0.45$ $H_1: \mu > 0.45$	0.000 324	10%
(d)	20.9	$H_0: \mu = 12.7$ $H_1: \mu \neq 12.7$	10	1%
(e)	25.8	$H_0: \mu = 30.6$ $H_1: \mu < 30.6$	8.4	5%

2 Samples of varying size are taken from populations of known variance. Test each pair of hypotheses at the level indicated.

	n	x	Hypotheses	Variance	Significance level
(a)	50	21.6	$H_0: \mu = 22.3$ $H_1: \mu \neq 22.3$	7.3	5%
(b)	100	4.87	$H_0: \mu = 5.24$ $H_1: \mu < 5.24$	2.24	1%
(c)	80	0.045	$H_0: \mu = 0.04$ $H_1: \mu > 0.04$	4.45×10^{-4}	2%
(d)	200	785	$H_0: \mu = 792$ $H_1: \mu \neq 792$	2332	3%
(e)	120	51.5	$H_0: \mu = 48.9$ $H_1: \mu > 48.9$	237	4%

3 A packet of plastic ring-reinforcements for file paper is supposed to contain 250 rings. In fact, the number of rings (R) in a packet is normally distributed as $R = N(250, 1.21)$. A single packet, chosen at random, is found to contain 248 rings. Does this mean, at a 5% significance level, that the mean is less than 250?

4 Bananas are delivered to a supermarket in large boxes containing 200 fruits each. It is known that, on arrival at the supermarket, 1% of the fruits are damaged. From a sample of 50 bananas, two are found to be damaged. Test at the 5% level to see if this implies that the proportion of damaged fruit is greater than 1%.

*5 A machine is used to produce cylindrical rods whose diameter, D cm, is normally distributed as $D = N(1.84, 0.0072)$. Every three months the rods are checked to see if the machine cutting surfaces need to be replaced. One check on 50 rods showed a mean diameter of 1.86 cm. At a 5% level of significance, does this indicate that replacement is needed?

6 The head teacher of a large school reports at a parents meeting that the percentage of pupils playing truant has been reduced to 2.4%. Some weeks after the meeting, the staff feel that the level of truancy has increased, and they carry out a check on a random selection of 400 pupils. Fifteen are found to be absent without a valid excuse. Test the hypothesis, at a 10% significance level, that the truancy percentage has increased.

7 The company making Calfast calculators claims that six out of ten A level mathematics students use their calculators. In an A level examination room, where 84 candidates were writing a mathematics paper, 39 were using a Calfast calculator. Does this result, at the 1% significance level, imply that the company has overstated its claim?

8 Two random samples, each consisting of 100 household members, were asked about the number of hours a junior member of the house spent watching television each week. The mean and variance of the samples were as follows.

Sample 1: $X_1 = 6.99$ hours, $\sigma_1 = 1.4$
Sample 2: $X_2 = 7.5$ hours, $\sigma_2 = 1.5$

Test at the 1% level to see if the mean of sample 2 is significantly higher than that of sample 1.

9 A firm making cricket balls labels them as having a mass of 156 g. Research at the factory shows that the masses (M g) of the balls are normally distributed as $M = N(156, 34.6)$.

(a) A sample of 120 balls shows a mean mass of 154.8 g. Test this at the 5% level to see if this value implies a different population mean.
(b) A second sample of 150 balls has a mean of 156.9 g. Does this indicate that the population mean should be greater than 156 g? Test this also at the 5% level.

10 Two firms make hockey sticks whose nominal mass is 21 oz. This mass in both populations is normally distributed. A random sample is taken from each population with the following results.

$S_1: n_1 = 50, \quad x_1 = 19.9 \text{ oz}, \quad \sigma_1^2 = 38$
$S_2: n_2 = 80, \quad x_2 = 22.08 \text{ oz}, \quad \sigma_2^2 = 43$

Test at the 5% level to see if there is a significant difference in the population means μ_1 and μ_2.
At the 10% level of significance, does the mean of S_1 indicate that the mean of the first firm's sticks is lower than 21 oz?

11 A population is normally distributed with mean μ_1 and variance 32.4, and a random sample of 75 from it has mean $\bar{x}_1 = 38$. A second normally distributed population has mean μ_2 and variance 40.7, and a random sample of size 60 from it has mean $\bar{x}_2 = 36.5$. Test at the 10% level to see if there is a significant difference in the two population means μ_1 and μ_2.

12 A schoolgirl buys a new battery for her quartz movement wristwatch. The jeweller assures her that the average life of the battery, under normal use, is 18 months. The girl, after corresponding with the makers of the battery, finds that the battery life, L months, is a normal variable $L = N(18, 12.96)$. This battery failed after only 11.8 months. Does this mean that makers' claim of an average 18 month life is not justified? Test at the 5% level.
Naturally, the girl complained to the makers and, as a result, they tested a sample of 200 batteries. The average life of this sample was 17.5 months. Should the makers revise their claim?

13 The figures shown below represent the results of six throws by a javelin thrower in the final Grand Prix meeting of a season. They are in metres correct to one decimal place.

$$87.4 \quad 81.6 \quad 83.5 \quad 86.8 \quad 88.2 \quad 80.7$$

The average throw made by this athlete, up to but not including this meeting, cannot be established because a careless official has lost the records. But it is thought to be 82.4 m. Calculate an unbiased estimate for the population variance of all the throws made by this athlete this season, then use the t-distribution to test whether the average of his throws at this meeting implies that the 'remembered' average is too low. Show your null and alternative hypotheses clearly.

14 Towards the end of an academic year, a mathematics teacher loses his mark book, which contains all the test results for his pupils during the year. He gives one A level class, whose average mark he thinks he remembers as 61%, a test just before the end-of-year examinations. The 12 pupils' marks, as percentages, are

54   57   62   58   60   67   61   55   56   49   58   59

Test at the 5% significance level to see whether the mean is less than 61.

7.2   Type I and Type II errors

Basic

1   A normal population has an unknown mean μ and variance 25. A sample of size 50 is taken and, as a guess, it is suggested that the population mean is 50. Two hypotheses are put forward to test this: H_0: $\mu = 50$, H_1: $\mu > 50$.
It is agreed that H_0 will be accepted if the sample mean, $\bar{X}$, is not more than 51.5.

(a)   Find the probability of a Type I error.
(b)   If the probability of a Type I error was laid down as 0.04, what would the value of 51.5 for the sample mean change to?

2   A spinner to be used at a village fete is pictured in the diagram. Before the fete, the spinner is tested to see if it is fair. The arrow is spun 108 times and the fete committee have decided that, if it stops in the red sector between 6 and 12 times inclusive, then it is fair; otherwise it must be rejected. Use a normal approximation to find the probability of rejecting a fair spinner.

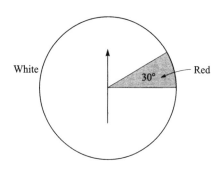

3   A set of 20 cards is known to contain some aces, and two hypotheses are made about the number of aces: H_0: there are 10 aces, H_1: there are 5 aces. To test this, four cards are removed at random, without replacement. If three or four of these are aces, H_0 is accepted; otherwise H_1 is accepted.
Find the Type I and Type II error probabilities.

4   The crv X is defined as $X = N(\mu, 25)$, and an estimated value for μ is 66. Two hypotheses are put forward: H_0: $\mu = 66$, H_1: $\mu > 66$.
If the probability of a Type I error is 0.2, find the critical region in which X will be rejected. What is the probability of a Type II error if in fact $\mu = 70$?

5 The crv X is described by two hypotheses:

$$H_0: f(x) = \begin{cases} \frac{1}{2}x & 0 < x < 2 \\ 0 & \text{otherwise} \end{cases} \qquad H_1: f(x) = \begin{cases} \frac{1}{10}(2x + 3) & 0 < x < 2 \\ 0 & \text{otherwise} \end{cases}$$

If $X = k$, where $0 < k < 2$, H_0 is accepted; otherwise H_1 is accepted.

(a) Find the probability of a Type I error for:
   (i) $k = \frac{3}{4}$ (ii) $k = 1$ (iii) $k = \frac{3}{2}$
(b) If H_1 were correct, find the probability of a Type II error in the cases where:
   (i) $k = \frac{7}{20}$ (ii) $k = \frac{13}{20}$ (iii) $k = \frac{7}{5}$

6 A normal variable X is defined as $X = N(\mu, 30)$. An informed guess suggests that its mean is 42. Two hypotheses are proposed:
$H_0: \mu = 42$, $H_1: \mu < 42$.
A sample of size 100 is used to test these hypotheses and it is agreed that H_0 will only be accepted if $\bar{X}$, the sample mean, is not less than 40.8. Find the probability of a Type I error.

7 A die is suspected of being biased towards showing a six. It is rolled 90 times and would be accepted as fair if a six were seen between 12 and 18 times inclusive; otherwise it is rejected. What is the probability that it is fair but is rejected as being biased?

8 X is a crv such that $X = N(\mu, 64)$. Two hypotheses about X are

$$H_0: \mu = 50, \qquad H_1: \mu > 50$$

A test is to be devised at the 5% significance level. The test statistic is the mean of a random sample of 30 observations of X.

(a) Find, to three decimal places the critical region of the test.
(b) Find the probability of a Type II error in the cases where:
   (i) $\mu = 51$ (ii) $\mu = 53$ (iii) $\mu = 55$

*9 A coin is suspected of being biased towards heads. The null hypothesis is $H_0: p(\text{heads}) = 0.5$, and the alternative hypothesis is $H_1: p(\text{heads}) > 0.5$. The coin is tossed 50 times and H_0 is to be accepted if the number of heads is less than 30.
Show that the significance level is just more than 10%.

10 A bag contains 20 cards, on each of which is written a different number, some of which are prime. Two hypotheses are made about the number, N, of primes in the bag. $H_0: N = 14$, $H_1: N = 12$. Three cards are drawn at random, without replacement. H_0 is accepted if the number of primes seen is two or three; otherwise H_1 is accepted. Find the probability of (a) a Type I error, (b) a Type II error.

11 The random variable X has a normal distribution such that $X = N(\mu, 200)$. Two hypotheses about μ are H_0: $\mu = 160$, H_1: $\mu \neq 160$. A single reading of X is taken and H_0 is to be accepted if $142 < \mu < 175$. What is the probability of a Type I error?

12 The pdf of the crv X is described by two hypotheses.

$$H_0: f(x) = \begin{cases} \frac{1}{4}(2x+1) & 1 < x < 2 \\ 0 & \text{otherwise} \end{cases} \qquad H_1: f(x) = \begin{cases} \frac{2}{7}(3x-1) & 1 < x < 2 \\ 0 & \text{otherwise} \end{cases}$$

H_0 will be accepted as long as $X < k$, where $k = \frac{8}{5}$; otherwise H_1 is to be accepted. Calculate:

(a) the probability of a Type I error,
(b) the probability of a Type II error, if k remains at $\frac{8}{5}$.

Intermediate

1 Two hypotheses about the pdf of a crv X are:

$$H_0: f(x) = \begin{cases} \frac{2}{3}(1+x) & 0 < x < 1 \\ 0 & \text{otherwise} \end{cases} \qquad H_1: f(x) = \begin{cases} 3x^2 & 0 < x < 1 \\ 0 & \text{otherwise} \end{cases}$$

A single reading of X is taken. If this value is less than k, where $0 < k < 1$, then we accept H_0 but, for any other value of X, H_1 is accepted.

(a) Find the probability of a Type I error for the following values of k.
   (i) $k = 0.5$ (ii) $k = 0.2$ (iii) $k = 0.75$
(b) Find the value of k in the following cases.
   (i) p(Type I error) $= 0.4$ (ii) p(Type I error) $= 0.6$
   (iii) p(Type I error) $= 0.8$
(c) For the following values of k, find the probability of a Type II error.
   (i) $k = 0.3$ (ii) $k = 0.4$ (iii) $k = 0.7$
(d) Find the value of k in the following cases.
   (i) p(Type II error) $= 0.3$ (ii) p(Type II error) $= 0.5$
   (iii) p(Type II error) $= 0.9$

2 A selection panel chooses 20 finalists in a beauty competition. Disappointed parents complain that the panel is biased towards blondes. To test this, the following procedure is carried out. First, two members of the public, who did not witness the competition, are asked to state null and alternative hypotheses respectively. Their choices are H_0: $N = 12$ blondes, H_1: $N = 10$ blondes. Second, the names of five finalists are chosen at random by one of the disappointed parents.

It is agreed that if four or five of the chosen contestants are blondes, then H_0 will be accepted, but not otherwise.
What is the probability that:
(a) H_0 is true but H_1 is accepted,
(b) H_1 is true but H_0 is accepted?

3 Two functions are put forward to describe the crv X.

$$H_0: f(x) = \begin{cases} \frac{6}{17}x & 0 < x < 1 \\ \frac{6}{17}x^2 & 1 < x < 2 \\ 0 & \text{otherwise} \end{cases} \qquad H_1: f(x) = \begin{cases} \frac{3}{8}x^2 & 0 < x < 2 \\ 0 & \text{otherwise} \end{cases}$$

To test this, it is decided that H_0 will only be accepted if $X < \frac{5}{4}$. Find:
(a) $p(H_0$ is true but H_1 is accepted),
(b) $p(H_1$ is true but H_0 is accepted).

4 The crv X is to be specified by one of two hypotheses for its pdf.

$$H_0: f(x) = \begin{cases} \frac{2}{3}(x+1) & 0 < x < 1 \\ 0 & \text{otherwise} \end{cases} \qquad H_1: f(x) = \begin{cases} \frac{3}{4}(x^2+1) & 0 < x < 1 \\ 0 & \text{otherwise} \end{cases}$$

(a) One observation of X is recorded as the number k. H_0 will be accepted if $k < \frac{2}{3}$; otherwise H_1 will be accepted. Find the significance level of this test.
(b) If the significance level were 10%, what would be the value of k?
(c) For the value of $k = 0.8$, find p(Type II error) for this test.

5 The following two hypotheses are made about the crv X.

$$H_0: f(x) = \begin{cases} \frac{1}{2}x & 0 < x < 2 \\ 0 & \text{otherwise} \end{cases} \qquad H_1: f(x) = \begin{cases} \frac{1}{2} & 0 < x < 2 \\ 0 & \text{otherwise} \end{cases}$$

One observation of X is made, and this can be used in two ways. Calculate the required quantity in each case.
(a) H_0 is to be accepted if $X < y$ where $y = 1$. Find p(Type I error),
(b) H_0 will be accepted if $X < z$, where z is such that p(Type I error)$=4p$(Type II error). Find the value of z.

6 During a recent tour of Stumpsland, the England cricket captain called 'heads' for the toss in each of the five Test matches and several one-day internationals which were played. He only won the toss once and so suggested that the coin used (the same coin each time) should be tested for bias towards tails.
The procedure was to toss the coin 200 times, and if fewer than 115 tails were observed, then H_0, $\{H_0: p(\text{tail}) = \frac{1}{2}\}$ was to be accepted; otherwise H_1, $\{H_1: p(\text{tail}) > \frac{1}{2}\}$ would be accepted.
(a) Find (i) $p(H_1$ accepted $\mid H_0$ true), (ii) $p(H_0$ accepted $\mid H_1$ true), if $p(\text{tails}) = 0.65$.

(b) How would the value of 115 tails have to be modified if the significance level of the test had been set at 5%?

7 A die is suspected of being biased towards showing a six. To test this, it is tossed 40 times and two hypotheses are advanced: H_0: $p(6) = \frac{1}{6}$, H_1: $p(6) > \frac{1}{6}$.
It is laid down that, if a six is seen fewer than 10 times, H_0 is accepted; otherwise H_1 is accepted.

(a) Find $p(H_1$ is accepted $\mid H_0$ is true),
(b) If the significance level of the test were 5%, what would be the maximum number of sixes allowed for H_0 to be accepted?

8 A bag contains 50 discs, coloured either red or white. Two hypotheses are made about the numbers of each colour:
H_0: there are 40 red and 10 white discs,
H_1, There are equal numbers of each colour.
To test H_0 against H_1, four discs are removed at random and without replacement. H_0 is accepted if all four discs are red; otherwise H_1 is accepted.

(a) Find $p(H_1$ is accepted $\mid H_0$ is true).
(b) Again find $p(H_1$ is accepted $\mid H_0$ is true), if this time, the selection is done with replacement.
(c) Find $p(H_0$ is accepted $\mid H_1$ is true), selection without replacement.

9 A supermarket sells easy-cook, long grain rice in bags whose mass is normally distributed as $X = N(500, 16)$. The mass of rice in a bag is regularly monitored, and one day a technician states that the mass in a bag is significantly more than 500 g. A sample of 20 bags gives a mean mass $\bar{X}$ g, and two hypotheses are made about the mean mass:
H_0: $\mu = 500$ g, H_1: $\mu > 500$ g. H_0 is to be accepted if $\bar{X}$ is less than a value M when the significance level of this test is 10%. Find the value of M to the nearest whole number.

10 A normal random variable X is described as $X = N(\mu, 150)$. Two hypotheses are made regarding μ: H_0: $\mu = 100$, H_1: $\mu < 100$.
To test these statements, the sample mean, $\bar{X}$, of 25 random observations of X is calculated. If $\bar{X} > k$, where $k = 95$, then H_0 is to be accepted; otherwise H_1 will be accepted as correct.

(a) Given that H_0 is in fact correct, find the significance level of this test.
(b) Find the value of k if the significance level is 10%.

11 A cubical die has the numbers 1, 1, 3, 3, 6, 6 marked on its six faces. It is suspected that the die is biased towards showing a three. To test this, two hypotheses are made: H_0: $p(3) = \frac{1}{3}$, H_1: $p(3) > \frac{1}{3}$.
The die is rolled 100 times. If a three is seen fewer than 40 times, H_0 is accepted; otherwise H_1 is accepted.

(a) Show that the significance level of this test is between 9% and 10%.
(b) In the case where $\mu = 35$, the variance remaining unchanged, find the probability of a Type II error.

12 A group of nine baby hedgehogs from three different litters is being cared for at an animal rescue unit after losing their mothers on roads. It is not certain how many males and females are in the group and the rescue unit staff run a small sweepstake in aid of funds. The two most popular suggestions, before the issue is decided by a hedgehog expert, are:

H_0: there are six males and three females,
H_1: there are four males and five females.

One of the staff has some knowledge of statistics and devises the following test.
Four babies are selected at random. If three or four are males, H_0 is accepted; otherwise H_1 will be accepted.

(a) Find $p(H_0$ is true, but H_1 is accepted).
(b) Show that the power of this test is about 83.3%.

13 Two hypotheses are made about the pdf of the crv X.

$$H_0: f(x) = \begin{cases} \frac{3}{43}x^2 & 1 < x < 2 \\ \frac{6}{43}x & 2 < x < 4 \\ 0 & \text{otherwise} \end{cases} \qquad H_1: f(x) = \begin{cases} \frac{2}{21}(x+1) & 1 < x < 4 \\ 0 & \text{otherwise} \end{cases}$$

H_0 is to be accepted if $X < k$, where $k = 1.5$, otherwise H_1 is to be accepted.

(i) Find the probability of a Type I error,
(ii) find the probability of a Type II error, if k is set at 3.

*14 For the pdf of the crv X, two suggestions are given:

$$H_0: f(x) = \begin{cases} \frac{1}{6}(2x+1) & 0 < x < 2 \\ 0 & \text{otherwise} \end{cases} \qquad H_1: f(x) = \begin{cases} \frac{1}{2}x & 0 < x < 2 \\ 0 & \text{otherwise} \end{cases}$$

H_0 is to be accepted if $X > k$, where $k = 0.5$, otherwise H_1 will be accepted.

(a) Find $p(H_1$ is accepted $\mid H_0$ is true).
(b) Find $p(H_0$ is accepted $\mid H_1$ is true).

15 A normal random variable X is defined as $X = N(\mu, 0.09)$. A random sample of 30 observations of X is taken, and its mean, $\bar{X}$, is calculated. Two hypotheses are made about μ: H_0: $\mu = 0.8$, H_1: $\mu > 0.8$. It is agreed that if $\bar{X} < k$, where $k > 0.8$, then H_0 is to be accepted; otherwise H_1 is to be accepted.
Find the value of k if the significance level of the test is (a) 1%, (b) 5%.

Advanced

1   The crv X is defined as $X = N(\mu, 45)$. Two hypotheses about μ are
    offered: H_0: $\mu = 80$, H_1: $\mu \neq 80$. The Type I error for this situation is set
    at 5%.

    (a) Find the limits of acceptance for μ (to three significant figures) for
        a sample size of (i) 15, (ii) 20, (iii) 40.
    (b) Find the Type II error in the cases where (i) $\mu = 79.1$,
        (ii) $\mu = 80.1$, (iii) $\mu = 82.3$, the sample size being 20. Use your
        answers to (a) to calculate these probabilities, also to three
        significant figures.

2   Two hypotheses about the pdf of the crv X are:

$$H_0 : f(x) = \begin{cases} \frac{1}{8}x & 0 < x < 4 \\ 0 & \text{otherwise} \end{cases} \qquad H_1: f(x) = \begin{cases} \frac{1}{20}(x+3) & 0 < x < 4 \\ 0 & \text{otherwise} \end{cases}$$

It is decided to accept H_0 if $X < k$, where $k = 3$; otherwise H_1 is
accepted. Find (a) the significance level of this test, (b) the power of the
test if the value of k is still taken as 3.

3   The well-known cardsharp, Mr Ace-up-sleeve, takes part in an
    experiment with some of his fellow gamblers. He is given eight identically
    backed cards, some of which are aces, while the rest are kings. He deals
    out three cards, 28 times, shuffling the pack each time. The number of
    times that three aces are seen in a hand is recorded. The fellow gamblers,
    who do not know the number of aces in the pack, make two hypotheses
    as to how many aces there are.

    H_0: there are four aces and four kings,

    H_1: there are six aces and two kings.

    It is agreed that, if three aces are seen fewer than five times, then H_0 will
    be accepted; otherwise H_1 is to be accepted (it is assumed that Mr Ace-
    up-sleeve does not cheat in this test).
    What is the probability that H_0 is correct, but it is rejected? Find also the
    probability that the pack contains six aces and two kings, but H_0 is
    accepted.

4   A hungry statistician finds that some food items in his refrigerator have
    gone off, so he decides to test the operating temperature of the machine.
    The manufacturers say that the temperature ought to have a normal
    distribution as $T = N(\mu, 1.44)$. The statistician makes two hypotheses:
    H_0: $\mu = 6°C$, H_1: $\mu > 6°C$, and he checks the temperature of 10 items.
    The mean, $\bar{X}$, of these temperatures is his test statistic, and he specifies
    1% for his Type I error.

Find the acceptance region for H_0. Find also the probability of a Type II error when $\mu = 8$, and find the power of this test.

5 For the pdf of the crv X, two suggestions are put forward:

$$H_0: f(x) = \begin{cases} \frac{1}{25}(x+1) & 1 < x < 2 \\ \dfrac{3(x^2 - 3)}{25} & 2 < x < 3 \\ 0 & \text{otherwise} \end{cases} \qquad H_1: f(x) = \begin{cases} \frac{1}{10}(3x - 1) & 1 < x < 3 \\ 0 & \text{otherwise} \end{cases}$$

As long as $X = k$, where $k < 2.5$, H_0 is accepted; otherwise H_1 is accepted. Find (a) $p(H_0$ is correct, but rejected), (b) the power of the test.

6 Two hypotheses are suggested for a random variable X:

$$H_0: f(x) = \begin{cases} \dfrac{6(x+2)}{91} & 0 < x < 1 \\ \dfrac{6(x^2 + 2)}{91} & 1 < x < 3 \\ 0 & \text{otherwise} \end{cases} \qquad H_1: f(x) = \begin{cases} \frac{1}{6}(2x - 1) & 0 < x < 3 \\ 0 & \text{otherwise} \end{cases}$$

H_0 is to be accepted provided $X < k$, where $1 < k < 3$ and the level of significance of this test is 5%; otherwise H_1 is accepted.
Show that, under these conditions, $k = 2.93$ to three significant figures.

*7 A bag contains 20 discs, some of which are coloured red, the remainder white. Two hypotheses about the number (N) of red discs are made: $H_0: N = 12$, $H_1 : N = 8$. Four discs are removed at random with replacement. H_0 is to be accepted if two or more are red; otherwise H_1 is to be accepted. Calculate to four significant figures: (a) p(Type I error), (b) p(Type II error), (c) the power of the test.

8 It is proposed to describe the pdf of a random variable X by one of two hypotheses.

$$H_0: f(x) = \begin{cases} C(e^x + 1) & 0 < x < 2 \\ 0 & \text{otherwise} \end{cases} \qquad H_1: f(x) = \begin{cases} C(x^2 + 1) & 0 < x < 2 \\ 0 & \text{otherwise} \end{cases}$$

H_0 will only be accepted if $X < k$, where $k = \frac{5}{4}$; otherwise H_1 will be accepted. Calculate the probability of a Type I error being made in this process. In the case where $k = 1.4$, calculate the probability of a Type II error.

9 A factory has two independent production lines, S and L, which produce the same item. Items from line S have lengths which are normally distributed with mean 300 mm and standard deviation 0.6 mm. Items from line L have lengths which are also normally distributed but with mean 301 mm and standard deviation 0.8 mm.

A quality inspector has a sample of six items from one of the lines which she thinks was S. She decides to measure the lengths of the items and attribute them to line S if the sample mean is less than k mm; otherwise she attributes them to line L.

(i) Find the value of k, correct to one decimal place, such that the significance level of her test is 5%.

(ii) Calculate the probability that the inspector wrongly attributes the six items to line S.

(iii) Do you consider the inspector's test to be reliable? Justify your answer. The inspector subsequently finds a sample of eight items, again thought to come from line S, and decides to use the value for k found in (i).

(iv) Calculate the probability that she correctly attributes these 8 items to line S.

[AEB]

Revision

1   The crv X is defined as $X = N(\mu, 12.4)$. Two hypotheses are made about μ: H_0: $\mu = 6.5$, H_1: $\mu > 6.5$. A sample of size 30 is tested, and H_0 is to be accepted if $\bar{X}$ is not greater than 7.5. Show that the significance level of this test is 6%.

2   A bag contains 10 balls, some of which are coloured black, the rest being white. Two suggestions are made about the number (N) of black balls: H_0: $N = 5$, H_1: $N = 3$. Three balls are removed from the bag at random and without replacement. H_0 will be accepted if all three are black; otherwise H_1 will be accepted. Calculate:

(a) the probability of a Type I error,
(b) the power of the test.

3   The crv X is normally distributed such that $X = N(\mu, 0.25)$. Regarding μ, we have two hypotheses: H_0: $\mu = 3.2$, H_1: $\mu > 3.2$. The mean of a random sample of 30 values of X is $\bar{X}$, and the significance level of this test is 10%.

(a) Find acceptance limits for μ.
(b) Find p(Type II error) when (i) $\mu = 3.29$, (ii) $\mu = 3.21$, (iii) $\mu = 3.09$.

4   A normal population has mean μ and variance 1. Two proposals are made for μ: H_0: $\mu = 40$, H_1: $\mu < 40$. A single reading is taken from the population. If it is more than 38.2, then H_0 is accepted but, for all other values, H_1 is accepted.

(a) Find (i) p(Type I error occurs), (ii) p(given that $\mu = 39$, that H_1 is correct, but H_0 is accepted).
(b) If, instead of a single reading, a random sample of 10 readings gave a sample mean of 39.4, what is now the probability of a Type II error, if $\mu = 40$?

5 A tetrahedral die marked 1, 2, 3, 4 is suspected of being biased in favour of showing a 'one'. Two hypotheses are made about this: H_0: p(one) $= 0.25$, H_1: p(one) > 0.25. The die is rolled 100 times. H_0 is to be accepted if the number of 'ones' observed is less than 31; otherwise H_1 is to be accepted. Find the probability that H_0 is true, but that the die is accepted as being biased.

*6 For the crv X, defined as $X = N(\mu, 16)$, the following two hypotheses are put forward: H_0: $\mu = 72$, H_1: $\mu < 72$. A single sample of 15 observations is made, and its mean, $\bar{X}$, is calculated. H_0 will be accepted if $\bar{X} > k$, otherwise H_1 is accepted. Find the value of k if the significance level of this test is 10%.

7 Two hypotheses are put forward to define the mean of the crv X:

$$H_0: \mu = 5.8, \quad H_1: \mu \neq 5.8. \quad X \text{ is defined as } X = N(\mu, 1.21).$$

H_0 will be accepted if the mean of a sample of 20 random observations of X is less than a number k. k is defined as giving the test a 4% level of significance. Calculate the value of k if H_0 is to be accepted.

8 The random variable X is defined as $X = N(\mu, 5.8)$. Two hypotheses about μ are considered: H_0: $\mu = 18.5$, H_1: $\mu < 18.5$. H_0 is to be accepted if $\bar{X}$, the mean of a sample of 25 random observations of X, is greater than 17.9. Find the probability of a Type I error. In the case that $\mu = 18.3$, find the probability of a Type II error.

9 The crv X has a proposed pdf f(x), where

$$f(x) = \begin{cases} \frac{2}{3}x & 1 < x < 2 \\ 0 & \text{otherwise} \end{cases}$$

This pdf is to be accepted for X, in preference to any other proposal, if $X < k$, where $k = 1.6$. Find the probability that f(x) is rejected, even though $k < 1.6$.

10 Two pdf's are proposed for the crv X.

$$H_0: f(x) = \begin{cases} \frac{2}{15}(x+1) & 0 < x < 3 \\ 0 & \text{otherwise} \end{cases} \qquad H_1: f(x) \begin{cases} \dfrac{x^2}{9} & 0 < x < 3 \\ 0 & \text{otherwise} \end{cases}$$

H_0 will be rejected and H_1 accepted if $X > k$, where $k = 2$. Find:

(a) the significance level of this test,
(b) the probability that H_1 is true, but that H_0 is accepted.

8

χ^2 tests

Degrees of freedom, goodness of fit, contingency tables, Yates correction

$$\chi^2 \text{ test}, \quad \text{test statistic is } \sum \frac{(f_o - f_e)^2}{f_e}$$

Basic

1 Use χ^2 tables to finds the following probabilities.
   (a) (i) $p(\chi_2^2 < 9.210)$ (ii) $p(\chi_5^2 < 1.145)$ (iii) $p(\chi_{10}^2 < 18.31)$
        (iv) $p(\chi_2^2 < 13.82)$ (v) $p(\chi_7^2 < 16.01)$
   (b) (i) $p(\chi_1^2 > 3.841)$ (ii) $p(\chi_5^2 > 15.09)$ (iii) $p(\chi_{10}^2 > 29.59)$
        (iv) $p(\chi_2^2 > 0.1026)$ (v) $p(\chi_6^2 > 14.45)$

2 In each case, find the value of x.
   (a) (i) $p(\chi_3^2 < x) = 0.990$ (ii) $p(\chi_5^2 < x) = 0.025$
        (iii) $p(\chi_8^2 < x) = 0.950$ (iv) $p(\chi_{10}^2 < x) = 0.050$
        (v) $p(\chi_{50}^2 < x) = 0.975$
   (b) (i) $p(\chi_2^2 > x) = 0.950$ (ii) $p(\chi_6^2 > x) = 0.010$
        (iii) $p(\chi_5^2 > x) = 0.975$ (iv) $p(\chi_{15}^2 > x) = 0.990$
        (v) $p(\chi_{12}^2 > x) = 0.050$

3 Find the following probabilities.
   (a) $p(9.348 < \chi_3^2 < 16.27)$
   (b) $p(0.05064 < \chi_2^2 < 0.1026)$
   (c) $p(31.53 < \chi_{18}^2 < 34.81)$
   (d) $p(2.088 < \chi_9^2 < 2.700)$
   (e) $p(0.5543 < \chi_5^2 < 1.145)$

4 The random variable X has a χ^2 distribution with d degrees of freedom. If $p(X > x) = \alpha$, find the value of α in each case:
   (a) $d = 5, x = 15.09$ (b) $d = 2, x = 7.378$
   (c) $d = 9, x = 27.88$ (d) $d = 15, x = 27.49$
   (e) $d = 6, x = 12.59$

*5 A cubical die is thrown 240 times, and the frequencies of the scores noted, with the following results.

Score	1	2	3	4	5	6
Frequency	35	27	37	55	51	35

Use a χ^2 test, at the 5% level, to see if the die is fair.

6 A tetrahedral die is rolled 200 times and the frequencies of the scores are noted below.

Score	1	2	3	4
Frequency	41	65	52	42

Show that, using a χ^2 test at the 5% level, the die can be regarded as fair. Would it be accepted at the 10% level?

7 The proportion of left-handed people in an urban area of Britain was found in a recent survey to be 12%. A random test group of 2000 from this population contained 215 left-handed people. Does this agree with the survey result if it is tested using a χ^2 test at the 10% level?

8 In each of the following sets of figures, the null hypothesis is that every value of x is equally likely to appear. Test each set at the level requested to see if the observed frequencies agree with this.

(a)
x	5	6	7	8	9	10	1%
f	11	24	35	33	26	21	

(b)
x	0	1	2	3	4	5	6	7	8	9	5%
f	29	43	38	21	28	41	49	32	37	32	

(c)
x	0	1	2	3	10%
f	19	31	21	29	

(d)
x	A	B	C	D	E	F	G	5%
f	180	225	212	187	185	206	205	

(e)
x	0	1	2	3	4	5	1%
f	37	53	42	65	67	36	

9 For each part of this question, test the null hypothesis using a χ^2 test at the required significance level. X is a drv in each case.

(a)
x	0	1	2	3	≥ 4	$H_0: X = \text{Po}(2)$ at 10%
f	8	33	35	15	9	

(b)
x	0	1	2	3	4	≥ 5	$H_0: X = \text{Po}(4)$ at 5%
f	8	12	22	45	30	83	

(c)

x	0	1	2	3	≥ 4
f	258	430	395	180	137

$H_0: X = Po(1.6)$ at 1%

(d)

x	0	1	2	3
f	15	48	26	11

$H_0: X = B(3, 0.4)$, at 10%

(e)

x	0	1	2	3	4	5
f	10	18	45	23	18	6

$H_0: X = B(5, 0.44)$, at 2.5%

(f)

x	0	1	2	3	4
f	33	85	168	96	18

$H_0: X = B(4, 0.5)$, at 10%

10

x	0	1	2	3	4
f	51	70	29	8	2

It is thought that the figures above are observations of a drv X which fits the binomial distribution $X = B(4, 0.2)$. Test this hypothesis at the 1% significance level.

11 Apply a χ^2 test at the 5% significance level to the 2×2 contingency tables shown below, to see if the variables P and Q are independent.

(a)

	Q_1	Q_2
P_1	3	17
P_2	15	25

(b)

	Q_1	Q_2
P_1	15	25
P_2	5	55

(c)

	Q_1	Q_2
P_1	10	20
P_2	40	30

(d)

	Q_1	Q_2
P_1	16	28
P_2	41	35

(e)

	Q_1	Q_2
P_1	34	56
P_2	49	61

Intermediate

1 A calculator is used to generate 700 random numbers. The final digit of each number is recorded and this gives the frequency table shown below.

Final digit	0	1	2	3	4	5	6	7	8	9
Frequency	59	74	80	61	58	77	88	91	55	57

Using a χ^2 test at the 1% significance level, determine whether the calculator is producing genuine random numbers.

*2 1400 students take a mathematics examination, with the following results:

Grade	A	B	C	D	E
Frequency	145	356	497	312	90

The national scores for this examination show that 12%, 24%, 35%, 21% and 8% of candidates achieve the grades A to E respectively. Test the above figures with a χ^2 test at the 2.5% level to see if they differ significantly from the national proportions.

3 The following results are from a scientific investigation, and it is suspected that they may fit a binomial distribution with $p = 0.4$.

x	0	1	2	3	4
Frequency	42	98	121	35	6

Test this hypothesis with a χ^2 test at the 10% level.

4 A committee interviews 150 applicants for jobs over a six-month period. Each candidate is graded according to the committee's estimate of performance. The results are shown in the table below.

Grade	A	B	C	D	E
Frequency	12	29	68	33	8

Past records show that the proportions of candidates given the grades A to E are 10%, 15%, 35%, 30% and 10% respectively. Test these figures with a χ^2 test at the 10% significance level to see if they show bias on the part of the committee.

5 A stationery firm manufactures blue, black, red and green felt-tip pens. Records of past sales show the proportions of colours sold annually are 46%, 38%, 13% and 3% respectively.
A large retail group sells 400 000 of these pens in a particular year, the numbers of each colour being 183 300, 152 850, 51 760 and 12 090 respectively. Using a χ^2 test, find whether these figures supply significant evidence, at the 5% level, for the manufacturers to consider changing the proportions of the colours made.

6 Each set of figures below is thought to fit a binomial distribution with p as shown. Apply a χ^2 test at the level requested to see if this is so.

(a)
x	0	1	2	3
f	92	66	37	5

$p = 0.25$, test at the 10% level

(b)
x	0	1	2	3	4
f	46	32	15	6	1

$p = 0.2$, test at the 5% level

(c)
x	0	1	2	3	4
f	8	64	84	37	7

$p = 0.5$, test at the 2.5% level

7 A cubical die is rolled 120 times, giving the results shown below.

Score	1	2	3	4	5	6
Frequency	17	20	15	16	16	36

Two observers make separate null hypotheses about the die:

(a) H_0: the die is fair,
(b) H_0: the results follow a binomial distribution with $p(6) = \frac{1}{4}$.

Test these results separately using a χ^2 test at the 1% level of significance.

8 Show that the figures in the first frequency distribution below fit the binomial distribution $B(4, \frac{1}{2})$ when tested at the 10% or lower level with a χ^2 test, but that the second group need testing at the 5% or lower level if they are to fit the same binomial distribution.

(a)
x	0	1	2	3	4
f	31	90	137	115	27

(b)
x	0	1	2	3	4
f	33	89	134	117	27

9 Show that the figures below will fit the binomial distribution $B(6, \frac{1}{2})$ at the 2.5% level but not at the 5% level of significance.

x	0	1	2	3	4	5	6
f	6	84	246	337	228	89	8

10 In the interests of economy, the number of telephone calls made from an office during the lunch hour is recorded over a period of 150 days. The results are below.

Calls/hour	0	1	2	3	4	≥ 5
Frequency	11	27	36	36	26	14

Show that these figures follow a Po(3) distribution, but not a Po(4) distribution, when tested at the 5% level of significance with a χ^2 test.

11 It is suspected that the figures below fit the Poisson distribution Po(5).

x	0	1	2	3	4	5	> 5
f	7	23	56	64	73	94	183

Test this idea, using a χ^2 test at the 1% significance level.

12 Fifty athletes, competing for places in an international relay squad are given intensive coaching for a period before selection. Every athlete improved his best time during this period, and the figures shown (F_1), record the overall improvements, in seconds, to the nearest 0.1 second (the race is a 4×100 relay).

Time	$0 < t \le 0.2$	$0.2 < t \le 0.4$	$0.4 < t \le 0.6$	$0.6 < t \le 0.8$	$0.8 < t \le 1.0$
F_1	16	12	10	7	5
F_2	21	14	5	5	5

The F_2 figures represent the average results of this coaching of successive groups of 50 over several years. Use a χ^2 test at the 10% level to test the hypothesis that there has been no significant improvement in the current year.

*13 A cubical die has its faces marked 1, 2, 2, 5, 5, 6, and it is suspected of being biased towards showing a five. It is thrown 300 times and a five is seen 125 times. Use a χ^2 test at the 1% level to check this assertion.

14 50 young people join a slimming club and, after three months, their losses (in kg) are recorded as shown in the table.

Loss, W	$0 < W \le 1$	$1 < W \le 2$	$2 < W \le 3$	$3 < W \le 4$	$4 < W \le 5$	$5 < W \le 6$	$6 < W$
Frequency	3	5	8	10	9	8	7

It is thought that these figures follow a normal distribution with mean 4 and variance 8. Test this assertion, using a χ^2 test at the 10% significance level.

Advanced

1   A cubical die is eccentrically loaded, so that the probabilities of the numbers (1 to 6) on its faces showing when the die is rolled are as described by the following rules, where X is the crv, 'the probability of a particular score being seen'.

$$F(X) = \begin{cases} 0 & x \le 0 \\ \dfrac{x^3}{27} & 0 \le x \le 3 \\ 1 & x \ge 3 \end{cases}$$

$p(1) = F(0.5), p(2) = F(1.0) - F(0.5)$

$p(3) = F(1.5) - F(1.0), p(4) = F(2.0) - F(1.5)$

$p(5) = F(2.5) - F(2.0), p(6) = F(3.0) - F(2.5)$

The die is rolled 200 times, with the following results.

Number seen	1	2	3	4	5	6
Frequency	3	11	21	42	59	64

Use a χ^2 test at the 1% level of significance to see if the figures fit the rules given above.

2 The lengths, L cm, to the nearest centimetre of 100 randomly chosen runner beans are shown in the table below.

L	$L \le 15$	$15 < L \le 19$	$19 < L \le 23$	$23 < L \le 27$	$L > 27$
Frequency	9	17	30	34	10

Estimate the mean and variance of this data from these figures, and then use a χ^2 test at the 5% level to see if they follow a normal distribution, using the calculated values.

*3 Over a period of 100 days, the number of hours a pianist spends practising each day is given by the figures below.

Hours	0	1	2	3	4	5	6	7	≥ 8
Days	4	5	8	12	19	22	15	9	6

Calculate the average number of hours of practice the pianist does per day. Use this average to calculate the frequencies of a comparable Poisson distribution. Use a χ^2 test at the 10% level of significance to see if his hours fit the Poisson distribution.

4 The contingency table below displays the observed frequencies P, Q, R and S.

	A	B
I	P	Q
II	R	S

Show that the expected frequencies for A/I and B/I, P_E and Q_E, respectively, are related as given below.

(a) $P_E + Q_E = \dfrac{(P+Q)^2 + (P+Q)(R+S)}{P+Q+R+S}$

(b) $P_E - Q_E = \dfrac{(P+Q)(P-Q+R-S)}{P+Q+R+S}$

5 An investigation into whether men or women enjoy driving a car yielded the following results. A total of 255 men and 245 women were interviewed.

	Men	Women
Enjoy driving	133	162
Do not enjoy driving	122	83

Test these figures with a χ^2 test at the 1% significance level to see if there is a significant difference between proportions of men and women drivers who enjoy driving.

6 An investigation into the preferences people have for red or white wine consisted of interviews with 1000 randomly chosen individuals from areas of east or west London, where they were asked to state their choice. The figures obtained are given in the following contingency table.

		\multicolumn{2}{c}{Preference}
		Red	White
Area	East	238	158
	West	304	300

Do these figures, tested at the 2.5% level by a χ^2 test, show that the preference for red or white wine depends on the area where the person lives?

7 In a large school, 120 pupils are taking A level mathematics. Three options are available to the pupils: pure mathematics, mechanics and statistics. The division of male and female pupils taking these options is shown in the contingency table. In this case, only one option is taken by each pupil.

	Male	Female
Pure mathematics	36	19
Mechanics	11	8
Statistics	19	27

Apply a χ^2 test at the 5% level of significance to see if the choice of option is dependent on the gender of the pupil.

8 A university shop sells sweatshirts which carry an embroidered university logo. The garments are made in red or blue, and the logo is designed to be within three different shapes: circular, triangular and square. The sales of 200 of these sweatshirts are recorded in the contingency table below.

	Red	Blue
Circle	39	56
Triangle	24	43
Square	17	21

Use a χ^2 test at the 10% level of significance to see if there is any difference between the sales of red and blue sweatshirts.

9 Observations were made at a bank to discover the pattern of use by males and females of the statement printer and the cash machine. The results are summarised in the following table.

	Statement printer	Cash machine	Total
Female	154	317	471
Male	92	340	432
Total	246	657	903

It is proposed to use a χ^2 test for independence. The expected frequency for females using the statement printer is 128.3, correct to 1 decimal place. Find the other expected frequencies.

Carry out the test at the 1% significance level and state your conclusion.

[UCLES]

Revision

1 Test each of the following sets of figures, which are values of the drv X, using a χ^2 test at the prescribed significance level, to see if they fit the suggested binomial distribution.

Readings   Level   Suggested distribution

(a)
x	0	1	2	3	4
f	6	19	30	39	6

5% $X = B(4, 0.6)$

(b)
x	0	1	2	3	4	5
f	5	14	16	34	6	5

5% $X = B(5, \frac{1}{2})$

(c)
x	0	1	2	3	4
f	5	23	11	8	3

10% $X = B(4, 0.35)$

2 Geraint is an enthusiastic rugby player who delights in making tackles. His coach records the number of tackles he makes during several five-minute periods during matches. From a large number of these periods, 50 are selected at random and the number of tackles in each period is shown in the table below.

Number of tackles	0	1	2	3	4	5	6	7
Frequency	1	2	3	18	16	5	3	2

Could these figures fit the binomial distribution B(7, 0.52) at the 10% level of significance if a χ^2 test is applied?

3 For each part of this question, test the null hypothesis using a χ^2 test at the required significance level. X is a drv in each case.

(a)

x	0	1	2	3
f	13	50	24	13

$H_0: X = B(3, 0.51)$, at 10%

(b)

x	0	1	2	3	4	5
f	18	49	41	7	3	2

$H_0: X = B(5, 0.3)$ at 5%

(c)

x	0	1	2	3	4
f	18	86	165	117	14

$H_0: X = B(4, \frac{1}{2})$ at 2.5%

(d)

x	0	1	2	3	≥ 4
f	6	9	18	24	43

$H_0: X = Po(3)$ at 10%

(e)

x	0	1	2	3	4	≥ 5
f	4	8	13	15	39	121

$H_0: X = Po(5)$ at 1%

(f)

x	0	1	2	3	≥ 4
f	200	350	400	220	230

$H_0: X = Po(2)$ at 5%

4 Could either of the sets of figures given below come from the Poisson distribution shown? X is a drv. Use a χ^2 test at the stated level.

(a)

x	0	1	2	≥ 3
f	15	43	31	11

$Po(1.4)$, test at the 2.5% level

(b)

x	0	1	2	3	4	≥ 5
f	34	60	61	28	11	6

$Po(2)$, test at the 10% level

5 Several countries decide to have a common currency in time for the AD3000 millennium. The new unit coin is called the ducat (to coin a name!), and its value fluctuates on the international money markets. Someone (perhaps a statistician!) remarks that, over periods of 50 randomly chosen days, the number of days on which the ducat's value differed from its launch value seemed to follow a Poisson distribution with mean 2. Here are the figures.

Differences (ducats)	0	1	2	3	4+
Frequency	8	20	32	23	17

Test this hypothesis with a χ^2 test at the 10% level.

6 In Britain, in the month of August, an observer can often see large numbers of shooting stars on a clear night. An enthusiastic observer recorded the number of these sightings which occurred in successive

10-minute intervals during several nights, producing the results shown below.

Number of stars per 10-minute interval:

$$0 \quad 1 \quad 2 \quad 3 \quad 4 \quad 5 \quad 6 \quad >6$$

Number of times this was observed:

$$15 \quad 31 \quad 33 \quad 28 \quad 18 \quad 7 \quad 6 \quad 2$$

Test these figures with a χ^2 test at the 5% level of significance to see if they fit a Poisson distribution with parameter 2.

7 X is a crv. 120 observations of X gave the following readings.

X	$0 < X \leq 5$	$5 < x \leq 10$	$10 < X \leq 15$	$15 < X \leq 20$	$20 < X \leq 25$	$25 < X$
f	12	21	23	25	28	11

The null hypothesis is H_0: $X = N(16.2, 50.41)$. Test this using a χ^2 test at the 10% level of significance.

8 The masses of 400 schoolboys are recorded below. They are in kilograms to the nearest kilogram.

Mass, M	≤ 54	$54 < M \leq 58$	$58 < M \leq 62$	$62 < M \leq 66$	$66 < M \leq 70$	$70 < M$
Frequency	61	65	68	70	78	58

Can these figures be represented by the normal variable $X = N(62, 44)$? Test with a χ^2 test at a 5% level of significance.

9 A cubical die is thrown 360 times and 72 sixes are recorded. Does this show that the die is biased towards the six? Use a χ^2 test at the 10% level.

10 A firm employs 248 staff, of whom 139 are men. According to an equal opportunities suggestion, equal numbers of men and women ought to be employed. Does a χ^2 test at the 5% level support the complaint that the firm is biased towards employing men?

*11 Test the figures in the contingency tables given to see whether the random variables P and Q are independent. Use a χ^2 test at the level indicated.

		X	Y				X	Y	
*(a)	P	18	2		(b)	P	8	16	
	Q	14	16	1%		Q	30	26	10%

9

Non-parametric tests

Sign tests, Wilcoxon and Mann–Whitney tests

Basic

1 The figures given below are random samples from different populations. Use a sign test to check the hypotheses stated about the median at the required significance level.

Sample	H_0	H_1	Level
(a) 7, 5, 7, 6, 8, 10, 4, 7	$m = 6$	$m > 6$	5%
(b) 28, 34, 38, 21, 43, 26, 25, 17, 26, 29	$m = 30$	$m < 30$	5%
(c) 75, 63, 81, 66, 71, 69, 59, 52, 58	$m = 55$	$m > 55$	5%
(d) 13, 7, 16, 15, 6, 4, 12, 18, 15, 14	$m = 11$	$m > 11$	10%
(e) 85, 83, 76, 92, 115, 88, 97, 78, 65, 100	$m = 100$	$m < 100$	10%

2 For each set of figures below, two hypotheses are made about the median. Check these using a sign test at the requested significance level.

Sample	H_0	H_1	Level
(a) 4, 12, 5, 12, 5, 6, 15, 7, 16, 17, 8, 9	$m = 12$	$m \neq 12$	10%
(b) 32, 47, 40, 42, 43, 48, 36, 39, 45, 51	$m = 50$	$m \neq 50$	3%
(c) 7, 8, 9, 11, 12, 15, 12	$m = 8$	$m \neq 8$	22%
(d) 68, 60, 94, 89, 77, 103, 100, 95	$m = 100$	$m \neq 100$	4%
(e) 2, 9, 7, 5, 12, 3, 9	$m = 10$	$m \neq 10$	8%

3 The sets of marks given below show the results of various classes in their mock A level and the examination. Use a sign test at the level indicated to see if the second set shows an improvement on the first

(a)
Candidate	1	2	3	4	5	6	7	8	
Mock	42	48	37	53	69	51	47	52	
A level	48	51	42	48	62	60	51	58	10%

(b)
Candidate	1	2	3	4	5	6	7	
Mock	74	52	43	48	55	61	58	
A level	81	56	42	53	58	66	64	10%

(c)

Candidate	1	2	3	4	5	6	7	8	9	10		
Mock	55	61	72	68	49	58	82	74	51	63		
A level	59	54	76	71	48	64	83	79	46	65		10%

(d)

Candidate	1	2	3	4	5	6	7	8				
Mock	32	41	65	58	72	68	45	52				
A level	48	50	61	63	78	70	49	55		5%		

(e)

Candidate	1	2	3	4	5	6	7	8	9	10	11	12	
Mock	48	61	54	38	43	75	40	58	25	83	71	64	
A level	53	65	56	41	39	78	36	61	27	80	74	68	5%

4 Find the critical region for the random variable X, for the five distributions listed, (i) for a one-tailed test, (ii) for a two-tailed test, at the 5% significance level in each case.

(a) $X = B(10, \frac{1}{2})$ (b) $X = B(7, \frac{1}{2})$,   (c) $X = B(8, \frac{1}{2})$
(d) $X = B(15, \frac{1}{2})$ (e) $X = B(9, \frac{1}{2})$

5 During the 1998 World Cup Football competition, the times in hours (to the nearest hour), spent watching football by 15 schoolchildren who ought to have been studying for their A level examinations were recorded by anxious parents as shown.

21   25   15   18   24   27   28   32   26   21   19   24   27   28   30

Do these figure support the hypothesis that the median number of hours was greater than 22? Use a sign test at the 10% level of significance.

6 Each set of figures shown below is a random selection from a separate population. The null and alternative hypotheses concerning the median are listed. Use a Wilcoxon signed-rank test to see if H_0 is to be accepted at the stated level.

(a) 12   9   8   11   10   14   10   7   15
    $H_0: m = 13$, $H_1: m < 13$,   1%
(b) 4   7   5   6   5   3   8   4   9
    $H_0: m = 7$, $H_1: m < 7$,   5%
(c) 24   27   33   35   29   32   31   33   28
    $H_0: m = 28$, $H_1: m \neq 28$,   5%
(d) 138   151   167   144   149   155   156   136   142   158
    $H_0: m = 141$, $H_1: m \neq 141$,   10%

7 Seven random observations are taken from a distribution whose median is thought to be 40. Use a Wilcoxon signed-rank test to see whether, at a 5% significance level, the median is not 40.

25   47   38   43   56   61   34

8 A firm making light bulbs asserts that the median life of their product is 500 hours. Use a Wilcoxon signed-rank test at the 2% level to see if the

following lifetimes of 10 randomly selected bulbs indicate a higher median.

> 570   490   580   550   500   610   440   550   600   540

9 Twelve students take a national numeracy test for which the median mark, out of 50, is 33. The results are shown below.

> 33   41   30   43   36   29   25   34   38   42   39   35

Apply a Wilcoxon signed-rank test at the 5% level to see if the median mark for this group of students is above the national value.

10 Each part of this question shows two unpaired sets of numbers. Use (i) the Wilcoxon rank-sum test and (ii) the Mann–Whitney U-test (5% significance level) to see if they come from populations with a common distribution.

(a)
A	15	17	23	31	37	42	43	45
B	19	25	28	34				

(b)
S_1	63	64	69	73	75				
S_2	59	66	70	71	74	76	78	84	86	100

(c)
P_1	1	11	15	17	30	34	35	38	43
P_2	3	5	9	12	19				

*11 The times, in minutes to the nearest minute, taken to solve a problem by two groups of university students are shown in the table.

G_1	18	19	24	25	27	29	30	32
G_2	20	21	22	23	31			

Apply (a) a Wilcoxon rank-sum test at the 10% level, (b) a Mann–Whitney U-test at the 5% level to test whether the times of the group G_2 are significantly better than those of group G_1.

12 Test each of the sets of figures below, using (i) a Wilcoxon rank-sum test at the 10% level, (ii) a Mann–Whitney U-test at the 5% significance level, to see if they come from populations having the same distribution.

(a)
X	52	84	93	107	118	152	205	
Y	63	106	135	168	190	230	275	308

(b)
L	3.1	4.5	7.5	11.4	17.2	19.9
M	6.8	13.6	24.2	26.7		

13 For each of the following groups of paired numbers, use the Wilcoxon matched-pairs signed-rank test to determine whether the second set differs from the first according to the stated hypotheses at the required level of significance.

(a)
$$S_1 \quad 4 \quad 5 \quad 4 \quad 7 \quad 6 \quad 7 \quad 8 \quad 5$$

H_0: they have the same median

$$S_2 \quad 6 \quad 8 \quad 5 \quad 6 \quad 9 \quad 11 \quad 8 \quad 4$$

H_1: S_2 has a greater median than S_1,   5%

(b)
| S_1 | 18 | 17 | 16 | 19 | 16 | 18 | 19 | H_0: medians are the same |
| S_2 | 17 | 20 | 20 | 24 | 17 | 22 | 25 | H_1: medians are different, 5% |

Intermediate

1   The sets of figures shown represent scores in a general knowledge test given to groups of pupils in neighbouring schools. The null hypothesis is that there is no difference between the results, and the alternative hypothesis states that the second result is always the better. Test these hypotheses using the Wilcoxon matched-pairs signed-rank test, at the requested significance level.

(a)
| S_1 | 34 | 42 | 51 | 44 | 46 | 32 | 48 | 63 | 44 | 38 | |
| S_2 | 37 | 45 | 54 | 42 | 48 | 34 | 51 | 60 | 46 | 40 | 10% |

(b)
| S_1 | 57 | 63 | 48 | 54 | 68 | 75 | 51 | 42 | 45 | 52 | |
| S_2 | 60 | 61 | 53 | 59 | 69 | 77 | 50 | 41 | 52 | 64 | 5% |

(c)
| S_1 | 67 | 72 | 64 | 68 | 76 | 84 | 65 | 53 | 62 | 70 | 75 | 80 | 74 | 49 | |
| S_2 | 69 | 75 | 62 | 71 | 77 | 79 | 61 | 50 | 68 | 77 | 79 | 84 | 72 | 78 | 54 | 1% |

(d)
| S_1 | 42 | 48 | 35 | 54 | 49 | 51 | 46 | 61 | 52 | 50 | 47 | 42 | 54 | 55 | 64 | |
| S_2 | 48 | 55 | 39 | 53 | 54 | 56 | 58 | 62 | 48 | 53 | 52 | 40 | 59 | 63 | 70 | 5% |

2   (a)   The figures below represent the scores made by two groups of people who enter a spelling competition. The groups consist of ten married couples, separated into husbands (H) and wives (W), who compete against each other. There are 80 words in the test and one mark is given for each correct answer.

| H | 57 | 63 | 48 | 54 | 68 | 75 | 51 | 42 | 45 | 52 |
| W | 60 | 61 | 53 | 59 | 69 | 77 | 50 | 41 | 52 | 64 |

Use a sign test, at the 5% significance level, to see if there is any difference in the performance of the two groups.

(b)   Some weeks later, the husbands demand a rematch, and this produces the scores shown from a similar test.

H	37	45	54	42	48	34	51	60	46	40
W	34	42	51	44	46	32	48	63	44	38

Apply the same test to these figures. Are the husbands significantly better this time?

3 A squad of 16 soldiers, on a two-week training course, is required to attempt an obstacle course at the start and at the end of the two weeks. The results show that 13 of the squad reduced the time taken to complete the tasks. Does this result agree with the hypothesis that the median time is less at the end of the course, when a sign test is applied at the 2% significance level?

4 A school sets a mathematics test to a group of prospective entrants. The median mark for this test over the past ten years has an average of 63%. For 10 of the entrants, the scores in the latest test are as follows:

$$58 \quad 69 \quad 65 \quad 74 \quad 70 \quad 85 \quad 81 \quad 72 \quad 64 \quad 76$$

Use a sign test to see if, at the 1% significance level, the marks indicate an increase in the median mark for the past 10 years.

5 The burning times, in hours to the nearest hour, of a dozen large church altar candles are shown below.

$$57 \quad 64 \quad 62 \quad 70 \quad 61 \quad 62 \quad 65 \quad 58 \quad 63 \quad 66 \quad 59 \quad 56$$

The median burning time is supposed to be 65 hours. Use a Wilcoxon signed-rank test at the 5% level to see if these figures suggest a lower median time.

6 Two small A level mathematics classes sit a common test with the following results.

Class 1	15	21	28	30	35	49	63
Class 2	19	32	47	54	58	65	

Use (a) a Wilcoxon rank-sum test, (b) a Mann–Whitney U-test at the 5% level of significance to see if the second set of marks has a higher median than the first.

7 Test the two sets of figures below, using a paired-sample sign test at the 5% level of significance, to see if the median of the second set is higher than that of the first.

$$11.4 \quad 13.5 \quad 7.1 \quad 12.9 \quad 16.7 \quad 14.5 \quad 19.1 \quad 22.8$$
$$12.6 \quad 9.4 \quad 6.8 \quad 15.7 \quad 18.2 \quad 16.4 \quad 17.6 \quad 23.6$$

8 Before travelling to the Olympic Games, each athlete is medically tested. One test consisted of measuring the time taken for each athlete's pulse to return to normal after stepping on and off a bench 20 times. The times are measured in seconds to the nearest second, and, for one particular group of competitors, the results are given below.

Men   26 27 29 32 33 34
Women   28 30 31 35

Test these figures, using (a) the Wilcoxon rank-sum test at the 10% level, (b) the Mann–Whitney U-test at the 5% level to see if the women's pulses returned to normal more quickly than did those of the men.

9 A television audience is requested to vote for one of two aspiring magicians, M_1 and M_2, who each perform the same eight tricks. The results, as marks out of 20, are given below

Trick	1	2	3	4	5	6	7	8
M_1	15	12	13	13	13	9	17	16
M_2	17	14	16	11	17	14	18	15

Test these figures with a Wilcoxon matched-pairs signed-rank test at the 5% level. Do they suggest that M_2 is seen as a better magician than M_1?

10 A football team, apart from its goalkeeper, divides its players into five defenders and five forwards. One of the forwards claims that they can run faster than the defenders. The manager decides to test this claim, and pairs off (using a random process) the players in 100 m sprints. The results, in seconds, were as follows.

Defender   11.6   11.7   11.8   12.0   11.8
Forward   11.5   11.8   11.6   11.7   11.4

The results are to be tested, at both the 5% and 10% significance levels, using a Wilcoxon matched-pairs signed-rank test. If the null hypothesis is that the median speeds are the same and the alternative hypothesis is that the speeds of the forwards have a lower median time, what conclusions are to be drawn?

11 After a club rugby match, the coach of the home side jokingly says to the opposition captain 'I think our eight forwards' masses come from a distribution which has a lower median than your seven three-quarters' (he must have been a statistician at some point!). The opposition captain takes this seriously (instead of just laughing), and so all these 15 players are weighed, in kilograms to one decimal place.

Home forwards   198   205   219   212   228   254   243   239
Opposition three-quarters 196   214   225   200   187   191   201

Use (a) a Wilcoxon rank-sum test and (b) a Mann–Whitney U-test, each at the 5% level, to decide whether the median mass of the three-quarters is higher than that of the forwards.

12 The leader of a slimming group, boasting about his team's performance, states that he is sure the group's median mass is now less than 72.5 kg (to one decimal place). The figures he produces to support his claim are:

Member	1	2	3	4	5	6	7	8
Mass	73.8	75.6	71.0	72.5	68.9	74.7	79.8	83.4

Test this assertion, using a Wilcoxon signed-rank test at the 5% significance level, to see if the median mass is indeed less than the stated value.

13 A statement on the label of a bottle of vitamin tablets says that each tablet contains 0.025 g of a certain vitamin. The Standards Office decides to test this claim. A Wilcoxon signed-rank test is to be used at a 10% significance level.
Eight tablets, randomly chosen from different bottles, are tested and the quantities of the vitamin concerned are measured, with the following results.

Mass (g)   0.027   0.026   0.025   0.021   0.026   0.025   0.028   0.023

Does the test show that the mass of vitamin in the sample comes from a distribution whose median is less than the claimed amount?

*14 For each set of paired figures below, use a paired-sample sign test, at the level indicated, to see if there is significant evidence that the second set shows an increase in the value of the median over the first.

(a)
A	163	158	154	161	157	166	172	159	160	155	
P	165	159	153	164	158	168	173	160	163	154	5%

*(b)
P	2.5	2.8	3.4	3.1	3.5	2.7	2.4	2.6	2.9	3.2	3.3	2.8	
Q	2.4	2.6	3.6	3.1	3.6	2.9	2.6	2.8	3.2	3.3	3.5	2.9	4%

15 A man who lives half-way between two large towns decides to buy a new television set for his family. He wants to buy one which is very commonly available and checks its price (in £) at seven retailers in one town and five in the other. His results are:

Town 1	299	275	280	305	278	285	284
Town 2	301	277	290	295	302	300	

One of the man's daughters, a promising A level statistics student, tests the figures using (a) a Wilcoxon rank-sum test and (b) a Mann–Whitney U-test at the 5% level of significance to see if prices in town 2 are higher than those in town 1. What will her conclusions be?

Advanced

1 (a) Seven promising young cricketers have the following batting averages half-way through a season. They are then given some intensive coaching by the national batting coach and their results for the rest of the season are compared with their results before coaching.

Batsman	1	2	3	4	5	6	7
Before coaching	36.9	43.2	28.7	31.5	38.6	40.4	42.1
After coaching	40.3	42.4	31.6	33.7	39.1	40.1	43.0

Use a Wilcoxon matched-pairs signed-rank test to see if there is significant evidence, at the 5% level, to conclude that the coaching had produced an improvement.

(b) If the final average of batsman number 6 had been 40.6, the others remaining as above, would this change your conclusion, at the same significance level?

2 The red squirrel, under pressure from the grey species, has declined in numbers in the UK. Eight separate sites were observed on two dates, a year apart, and the number of red squirrels seen recorded as below.

Site	1	2	3	4	5	6	7	8
First year	15	12	14	9	13	15	10	16
Second year	12	11	10	11	10	13	9	18

Use a Wilcoxon matched-pairs signed-rank test at the 10% significance level to see if there is significant evidence that the numbers of red squirrels at these sites have declined.

3 A trout farm, which rents out time to visitors, advertises that the median mass of fish caught is 500 g. Twenty fish are caught and weighed, the figures below being given in grams to the nearest gram.

$$570 \quad 580 \quad 530 \quad 490 \quad 550 \quad 440 \quad 610 \quad 460 \quad 640 \quad 520$$
$$590 \quad 600 \quad 500 \quad 560 \quad 515 \quad 610 \quad 480 \quad 580 \quad 630 \quad 510$$

Use a Wilcoxon signed-rank test at the 1% level of significance to test whether the farmer is underestimating the median mass.
If the farmer raises his estimate to 530 g, show that this makes the new null hypothesis acceptable at the 5%, but not at the 10% level.

4 Ten competitors take part in a clay-pigeon shoot. Ten clays are thrown up for each competitor and the number broken is recorded. A week later, after regular practice, the same ten competitors compete again with the same number of clays. The two sets of results are displayed below.

Competitor	1	2	3	4	5	6	7	8	9	10
First shoot	5	7	6	4	5	6	3	7	6	4
Second shoot	5	8	7	5	6	7	4	6	7	5

Use a paired-sample sign test at the 2% significance level to see if these results support the assertion that the second list showed there had been an improvement.

5   A market garden produces a specialist pumpkin fertiliser, Marrogro. A test is carried out by planting 10 seeds, equally spaced, in each of two adjacent rows in a vegetable garden, on the same day. One row is fertilised with Marrogro and the other with a non-specialised fertiliser (NSF). All the pumpkins are harvested on the same day and the masses of each pair are recorded below.

	1	2	3	4	5	6	7	8	9	10
Row 1, NSF	3.1	3.7	4.2	2.9	3.4	3.8	4.1	2.7	3.2	3.5
Row 2, Marrogro	3.4	3.6	4.4	3.1	3.7	3.9	4.0	2.5	3.5	3.8

Use a paired-sample sign test to check if these figures support the statement that the use of Marrogro produces heavier pumpkins. Test at the 20% level.

6   An A level paper, lasting 3 hours, consists of 15 questions, of which not more than 7 are to be attempted. The examiner setting the paper allows 21 minutes per question, to give candidates time to read and check questions.
Ten randomly selected students were timed for the same question from the paper and the following results were obtained.

Candidate	1	2	3	4	5	6	7	8	9	10
Time (minutes)	20.2	21.4	18.7	19.9	20.8	21	20.3	24.1	20.6	19.5

Use a Wilcoxon signed-rank test at the 10% level to determine whether the examiner is justified in reducing the time allowance per question.

7   (a) A work-study team is checking on the time taken to assemble part of an electronic device. The manufacturer bases the charge for this item on a median time of 183 seconds. Eight randomly selected employees were timed, with the following results, to the nearest second.

176   189   194   186   188   197   181   185

Carry out a Wilcoxon signed-rank test at the 5% significance level to see if the median time ought to be raised.

(b) One of the work-study team, checking his figures, finds that the time of 181 seconds should have been 191 seconds. Show that this would cause rejection of H_0 at the 5% level and that H_0 would only just be acceptable at the 4% level of significance.

*8 The makers of Buffalo beer carry out a survey of the sales of their product. The number of pints sold per week at ten randomly selected public houses (before and after an advertising campaign) were found to be as follows.

Public house	1	2	3	4	5	6	7	8	9	10
Before	74	63	95	88	82	79	90	97	53	69
After	72	66	98	100	87	83	95	106	50	74

(a) Is there significant evidence, at the 1% level, that the median number of pints sold increased after the campaign? Use a Wilcoxon matched-pairs signed-rank test.

(b) If a paired-sample sign test were used on these figures, show that the null hypothesis would be accepted at the 5% but not at the 6% significance level.

9 Two school rowing eights are tested over a special course to assist the coach to select one to represent the school in an important regatta. They each row, in as similar weather conditions as possible, over the course several times and their times, in minutes and seconds, are as follows.

$$E_1 \quad 6.38 \quad 7.04 \quad 7.10 \quad 7.08 \quad 6.51 \quad 6.49 \quad 7.06$$
$$E_2 \quad 6.43 \quad 7.05 \quad 6.41 \quad 6.48 \quad 6.50$$

To distinguish between these results, two tests are used to see if the median time for the second eight is lower than that for the first:

(a) the Wilcoxon rank-sum test,
(b) the Mann–Whitney U-test.

What conclusions are to be drawn from the results of these two tests?

Revision

1 The lengths, in centimetres to the nearest centimetre, of a sample of 12 cucumbers taken at random from a large crate are shown below.

$$38 \quad 37 \quad 35 \quad 41 \quad 32 \quad 36 \quad 37 \quad 40 \quad 38 \quad 42 \quad 40 \quad 39$$

Use a sign test to check whether the median length of the cucumbers is greater than 36 cm, at the 5% significance level.

2 The contents of a pot of honey have a nominal mass of 454 g. A check on six randomly selected pots gives the following masses, in grams to the nearest gram.

$$458 \quad 450 \quad 461 \quad 457 \quad 459 \quad 460$$

Use a Wilcoxon signed-rank test at the 5% significance level to see if the median mass differs from 454 g.

3 Packets of runner-bean seeds contain 50 seeds each. The company marketing the seeds claims that 80% of the seeds in each packet will germinate. A test of eight randomly chosen packets showed the following germination figures.

$$42 \quad 43 \quad 36 \quad 41 \quad 38 \quad 45 \quad 40 \quad 39$$

Check the company's assertion, using a Wilcoxon signed-rank test, to see if the null hypothesis is to be accepted at the 10% level of significance.

4 A small group of mathematics students take a test which is marked out of 10. Before the test, the teacher said she hoped the median mark would be more than 7. The results were:

$$6 \quad 7 \quad 9 \quad 6 \quad 8 \quad 8 \quad 10 \quad 9$$

At the 10% significance level, does the result of a Wilcoxon signed-rank test realise the teacher's wish?

5 An A level class of 12 students takes a test which is marked out of 100. The results are:

$$58 \quad 43 \quad 67 \quad 60 \quad 59 \quad 73 \quad 48 \quad 52 \quad 64 \quad 53 \quad 44 \quad 82$$

The median mark for the previous test taken by this class was 51. Do these marks justify the assertion that the new median is greater than 51? Use a sign test at the 5% significance level.

6 A company markets two kinds of toothpaste, Tartarless and Fluorplus. In the course of a survey, ten people, chosen at random, agreed to use both pastes and say which one was preferred. Eight of these preferred Tartarless. At what level of significance (to one significant figure) does this result show bias towards Tartarless paste, using a sign test?

7 Kitefly, a company making kites, tests the strength of one of its models by applying a force perpendicular to the centre of the main cross spar. The results below, of 14 tests on kites randomly chosen from stock, show the maximum force (in Newtons) supported before the spars broke.

$$57 \quad 42 \quad 37 \quad 51 \quad 48 \quad 34 \quad 52 \quad 44 \quad 55 \quad 50 \quad 46 \quad 55 \quad 41 \quad 53$$

The company claims the median breaking force is more than 43 N. Do the figures agree with this, using a sign test at the (a) 5%, (b) 10% significance level?

8 Many rugby matches are won because one of the teams has an accurate kicker of goals. A club coach tries to improve the actions of the principal kickers in each of its 10 teams. A professional coach is employed to observe and help the 10 members. Before and after this coaching, each of the ten players takes 10 kicks from the same 10 spots on the field and the number of successes is recorded below.

Player	1	2	3	4	5	6	7	8	9	10
Before	8	8	6	7	7	8	7	6	6	4
After	9	8	5	8	6	9	8	9	7	6

Test, at the 5% significance level, to see if there has been improvement after the coaching, using

(a) a Wilcoxon matched-pairs signed-rank test,
(b) a paired-sample sign test.

*9 Use a Wilcoxon matched-pairs signed-rank test at the 5% significance level to see if the second set of figures in the paired sets below has a significantly greater median than the first.

A	31	33	30	28	34	32	28	29	35	27
B	32	34	28	31	38	35	24	31	40	33

10 An entirely new vegetable fertiliser, N-Large, is claimed to produce larger (and sweeter!) peaches and nectarines. Nectarines from two adjacent trees in an orchard, one of which was treated with N-Large and the other with a standard fertiliser, are selected at random, picked and weighed, giving the results shown (in grams to the nearest gram).

N-Large	138	151	142	136	140	147
Standard	130	143	145	141	127	

Use a Mann–Whitney U-test at the 5% level to see whether the N-Large treatment produces significantly heavier fruit.

Answers and solutions

Chapter 1 Position and dispersion, display and description

1.1

Basic

1 (a) (i)

No. of sets	0	1	2	3	4
Frequency	2	15	8	4	1

    (ii) 1, 0.9
    (iii) 1.57

  (b) (i)

No. of children	0	1	2	3	4	5
Frequency		1	5	12	4	2	1

    (ii) 2, 1.58
    (iii) 2.16

  (c) (i)

No. of cars	0	1	2	3	4
Frequency	2	31	12	4	1

    (ii) 1, 0.758
    (iii) 1.42

  (d) (i)

Grade	A	B	C	D	E
Frequency	3	7	11	5	4

    (ii) C, C
    (iii) Not possible

2 (a) (i) 5   (ii) 3.95   (iii) 4.37
  (b) (i) J   (ii) K   (iii) Not possible
  (c) (i) 1   (ii) 1.19   (iii) 1.78

*3   Mean $= \frac{1}{100}(1 \times 17 + 2 \times 15 + \cdots$
$+ 6 \times 18) = \frac{355}{100} = 3.55$

Median $= 3 + \dfrac{50.5 - 48}{66 - 48} = 3.14$

4 (a) (i) 4, 4.11   (ii) 4.62   (iii) 1.38
  (b) (i) 6, 3.96   (ii) 4.76   (iii) 2.36
  (c) (i) 30, 25.9   (ii) 27.4   (iii) 8.90
  (d) (i) 5.8, 5.37   (ii) 5.52   (iii) 1.62

5 (a) $10 \leq S < 15$, $15 \leq S < 20$
  (b) 18.2, 7.93

6 (a) (i) 1.43   (ii) 1.43
  (b) (i) 2.58   (ii) 2.58
  (c) (i) 1.63   (ii) 1.63

7 (a) (i)

10.5–12.5	12.5–14.5	14.5–16.5
6	13	16

16.5–18.5	18.5–20.5	20.5–22.5
8	4	3

    (ii) 14.5–16.5, 14.5–16.5
    (iii) 15.5

  (b) (i)

5.5–7.5	7.5–8.5	8.5–9.5
4	4	13

9.5–10.5	10.5–11.5	11.5–15.5
8	7	4

    (ii) 8.5–9.5, 8.5–9.5
    (iii) 9.65

  (c) (i)

45.5–48.5	48.5–51.5	51.5–54.5
1	2	8

54.5–57.5	57.5–60.5	60.5–70.5
5	3	1

    (ii) 51.5–54.5, 51.5–54.5 (iii) 54.7

8   50–59, 50–59; mean $= 53.1$, $\sigma = 20.0$

*9   Mean $= \frac{1}{75}(2 \times 9 + 4 \times 22 + 6 \times 27$
$+ 8.5 \times 17) = 5.5$

Variance $= \sum \left(\dfrac{fx^2}{n} \right) - \mu^2$

$= \dfrac{1}{75}(4 \times 9 + 16 \times 22 + 36$
$\times 27 + 72.25 \times 17) - 5.5^2$
$= 4.26$

10 -0.769, -0.462

11 (a) (i) 4.10, 1.86
    (ii) 2.27, 3.36, 4.91
    (iii) 0.419, 0.174
  (b) (i) 5.56, 1.84
    (ii) 4.15, 5.43, 6.44
    (iii) -0.239, -0.118

12 Class 1   23, 9.80; Class 2   35, 10.5

Intermediate

1 $A = 28,\ B = 14$

2   (a)

75–85	85–95	95–105
15	17	13

105–115	115–130
4	1

   (b) 85–95, 85–95
   (c) 91.9 cm, 10.3 cm

3   (a) $A = 10$ (b) 133, 6.86

4   18

5   (a) 8.57, 16.8   (b) 10.2, 7.04

6   1.07

7   (a) £4920, £816
   (b) (i) £4650
      (ii) 13
      (iii) £4950
      (iv) 0.110

8   109.3 kg, 7.00 kg

9   (a) $P = 14,\ Q = 13$ (b) 25

10 (a) Mode = 9, median = 7.65
   (b) 7.95, 1.41
   (c) −0.745, 0.638

* 11 (a) $x - 2 + x + x^2 + (x + 1)^2 + 2x$
      $+ x + 1 = 60$
      $\Rightarrow 2x^2 + 7x - 60 = 0$
      $\Rightarrow (2x + 15)(x - 4) = 0$, so
      $x = -\frac{15}{2}$ or 4, i.e. $x = 4$
   (b) Frequency distribution is

x	0	1	2	3	4	5
f	2	4	16	25	8	5

   Mean $= \frac{1}{60}(0 \times 2 + 1 \times 4 + \cdots + 5 \times 5)$
      $= \frac{168}{60} = 2.8$

   $$SD = \sqrt{\begin{aligned}&[\tfrac{1}{60}(0 \times 2 + 1 \times 4 + 4 \times 16 + 9\\&\times 25 + 16 \times 8 + 25 \times 5) - 2.8^2]\end{aligned}}$$

      $= 1.12$

12 8, 12, 16, 16, 4; both dice are biased
   towards 5

13 (a) 26, 11.4
   (b) Curve, median = 27.1
   (c) −0.977 or −0.888

14 (a) 55.6, 18.4
   (b) 60–75, 45–60
   (c) IQR $= 68 - 45 = 23$

15 £115, 16.6%

Advanced

1 $A = 7,\ \bar{x} = 22,$ median $= 21$

2   (a)

0–9	10–19	20–29	30–39
1	2	4	6

40–49	50–59	60–69	70–79
15	12	10	6

80–89	90–99
3	1

   (b) 51.7, 18.4
   (c) 108.4, 36.8
   (d) 203, 5
   (e) $p = 1.09,\ q = 43.8$

3   £31.89, 170 s, 171 s

4   (a) £64.97
   (b) 12 min 16 s
   (c) 10 min 56 s
   (d) 4 min 5 s

5

80–82	82–84	84–86	86–88
1	4	8	9

88–90	90–92
4	4

   Mean = 86.5, SD = 2.63
   Second horse faster, but first more
   consistent

*6 No. of diamonds

0	1	2	3	4	5	6

   No. of areas

2	10	13	19	27	21	8

Cumulative frequency

| 2 | 12 | 25 | 44 | 71 | 92 | 100 |

$$\text{Median} = 3 + \frac{50.5 - 44}{71 - 44} = 3.24,$$

$$\text{LQ} = 2 + \frac{25.5 - 25}{44 - 25} = 2.03,$$

$$\text{UQ} = 4 + \frac{75.5 - 71}{92 - 71} = 4.21$$

(a) Quartile coefficient

$$= \frac{4.21 + 2.03 - 2 \times 3.2}{4.21 - 2.03} = -0.0734$$

(b) Pearson's coefficient

$$= \frac{3.54 - 4}{1.5} = -0.307$$

7 12.5, 25, 20, 12.5, 10, 20
Biased: mean $= 3.43$, SD $= 1.70$
Fair: mean $= 3.50$, SD $= 1.71$

8 (a) Median $= 261.7$, LQ $= 252.3$,
UQ $= 297.2$
(c) Positively skewed (Pearson's
coefficient $= 0.876$); possible
outliers, 220.6, 326.3
(d) Mean $= 272.0$, SD $= 35.3$
(e) (i) Reduced   (ii) No change

9 (ii) Median $= 5.90$, LQ $= 4.65$,
UQ $= 7.05$ (range 2.4)
(iii) Mean $= 6.57$
(iv) No: unlisted numbers; not every
household has a telephone.

Revision

1 (a) (i)

0	1	2	3	4	5
2	21	14	5	2	1

(ii) 1.71, 1.02
(b) (i)

0	1	2	3	4	5	6
6	10	7	4	1	1	1

(ii) 1.7, 1.46

2 (a)

x	0	1	2
f	7	10	8

(b) Median $= 0.7$
(c) Mean $= 1.04$

3 Mean $= 133.5$ kg, SD $= 10.9$,
Median $= 134$ kg

4 (a) 0.00447 (b) 4014 (c) 2.0005455

5 (a) 1.13, 1.29 (b) 0.670, -0.224

*6 (a) Mean $= \frac{1}{150}(2 \times 3 + 4 \times 5 + \cdots$
$+ 14 \times 19) = 10.4$

$$\text{Median} = 10 + 2 \times \frac{75.5 - 75}{131 - 75}$$
$$= 10.0$$

Mode $= 12$
(b) Variance $= \frac{1}{150}(4 \times 3 + 16 \times 5$
$+ 36 \times 11 + \cdots$
$+ 196 \times 19)$
$-10.4^2 = 7.45$

(c) LQ $= 8 + 2 \times \frac{37.75 - 35}{75 - 35} = 8.14$,

$$\text{UQ} = 10 + 2 \times \frac{113.25 - 75}{131 - 75} = 11.4$$

Quartile coefficient

$$= \frac{11.4 + 8.14 - 20}{11.4 - 8.14} = -0.141$$

7 (a) 56.1, 5.86   (b) -0.972

8 (a) (i) 19.0 (ii) 39.3
(iii) 17.0 (iv) 18
(b) (i) 3.28 (ii) 0.206
(iii) 3.04 (iv) 3.1

9 (a) 3.24   (b) 1.28   (c) -0.0698

10 (a) 42.9, 19.3   (b) 40.8

11 85%

12 Mean $= 12.5$; yes, a slight
overstatement

1.2

Basic

1 S 216°, M 72°, G 54°, B and E 18°;
£100 000

2 (a) A 40°, B 80°, C $=$ D $= 120°$
(b) (i) £102 000   (ii) £306 000

*3 $15x + 6x + x + 5x + 90 = 360 \Rightarrow 27x = 270 \Rightarrow x = 10$

	Angles	Money
Salaries	$15 \times 10 = 150°$	$£\dfrac{150}{360} \times 1\,800\,000 = £750\,000$
Buildings	$60°$	£300 000
Books	$10°$	£50 000
Food	$50°$	£250 000
Maintenance	$90°$	£450 000

6 (a) 5 | 8 9 9 9

    6 | 0 1 1 1 2 2 2 2

    6 | 3 3 3 4 4 4 4 4 5 5 5 5

    6 | 6 6 6 6 7 7

    Key: 6 | 3 means 63 g

  (b) 10 | 8 9 9

    11 | 0 0 1 1 1

    11 | 2 2 2 2 2 2 3 3 3 3

    11 | 4 4

    Key: 11 | 2 means 11.2 seconds

8 0 | 11 50 54

  1 |

  2 | 93

  3 |

  4 | 13 49 56 58

  5 | 09 28 64

  6 | 30

  7 | 17 61 98

  8 | 17 65 72

  9 | 62 64

9 (a) 1.22　(b) 0.841

  (c) 0.787　(d) −0.689

10 (a) 22–24, −0.225

  (b) 59–61, 0.0915

  (c) 14.4–14.65, 0.430

11 Class 1　mean = 23, SD = 9.80

  Class 2　mean = 35, SD = 10.5

12 (a) Mode = 6, Mean = 5.84

  (c) Median(table) = 5.47,

    Median(graph) = 5.6, IQR = 1.15

13 (a) 17.6, 18, 8, 26

  (b) 25.2, 23, 16, 36

  (c) 142, 143.5, 136.5, 148.5

14 (a) 3　(b) 8　(c) 21

15 (a) 70–90　(b) 115–135

  (c) 170–200　(d) 10 000–20 000

Intermediate

1

Model	1	2	3
(a)	$124°$	$136°$	$100°$
(b)	$124.9°$	$130.7°$	$104.4°$

Pie chart radii $R_2 : R_1 = 1.05 : 1$

2

	M_1	M_2	M_3	M_4	
G_1	$40°$	$130°$	$70°$	$120°$	$R_1 = 3$ cm
G_2	$66°$	$120°$	$78°$	$96°$	$R_3 = 3.87$ cm
G_3	$60°$	$90°$	$90°$	$120°$	$R_3 = 2.45$ cm

3 A　E 86°　J 70°　NA 122°　SA 82°

  B　E 79°　J 72°　NA 118°　SA 91°

  $R_A = 4$ cm, $R_B = 5.7$ cm

4 1, 83°; 2, 72°; 3, 99°; 4, 106°

6 Pie chart

  PM 108°　M 54°　S 72°　P 36°　C 90°

  Bar chart

    30%　　15%　20%　10%　25%

  (% of total)

7 1 | 4 9

  2 | 1 3 5

  3 | 3 4 4 6 7

  4 | 1 3 4 4 5 6 7 8 8

  5 | 1 2 4 5

  6 | 0 1

  Key: 3 | 6 means 0.36

  Class intervals: 0.1–0.19, 0.2–0.29 . . .

8 $x = 20$

March–May	June–August
63°	31°
3324	1636

September–November
46°
2427

December–February
220°
11 609

*9

```
4 | 95
5 | 06 11
5 | 39
5 | 54 57
5 | 60 63 66 66 68 71 73 74 75 76 78 79
5 | 80 80 82 83 83 84 84 85 86 87
  |            88 92 93 94 94 98
6 | 04 05 08
6 | 21 27 33
6 | 53
```

Class intervals: 48–49.9, 50–51.9,
52–53.9, ...
Key: 5 | 06 means 50.6
     6 | 27 means 62.7

10

Angles	S 216°	C 90°	PW 36°	RA 18°
Costs	£120 000	£50 000	£20 000	£10 000

11 (c) Median = 103,
IQR = 104.6–101.4 = 3.2

12 (a) S £540 000   V £342 000
M £180 000   A £18 000
(b) S 180°   V 114°
M 60°   A 6°

*13 (a) Angles:

$$News = \frac{5.5}{18} \times 360 = 110°$$

$$Phoenix = \frac{2.4}{18} \times 360 = 48°$$

$$Standard = \frac{5.8}{18} \times 360 = 116°$$

$$Monitor = \frac{4.3}{18} \times 360 = 86°$$

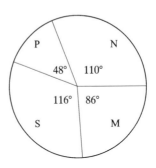

(b) Second chart angles: *News* 100°,
Phoenix 60°, *Standard* 120°,
Monitor 80°.
Changes: *News*,

$$First \ \frac{110}{360} \times 100 = 30.6\%,$$

$$Second \ \frac{100}{360} \times 100 = 27.8\%;$$

2.8% fall
Similarly, *Phoenix* 3.4% rise,
Standard 1.1% rise,
Monitor 2.2% fall

14 (a) 120°   (b) 150°   (c) 100 acres

16 (a) (ii) Mean = 108 km h^{-1},
median = 105, LQ = 98.6,
UQ = 114
(b) (ii) Mean = £1044.50,
median = £1017.11,
LQ = £914, UQ = £1144

17 (b) 50°, 115°, 94°, 65°, 36°
(c) Median = 31.5 minutes
(d) Mean = 34.4 minutes

Advanced

1 (a) 15%   22%   46%   17%
(b) $\mu_1 = 90$, $\sigma_1 = 44.602242$
$\mu_2 = 25$, $\sigma_2 = 12.389512$;
$\mu_1 = 3.6\mu_2$

2 (a) SS 660, L 300, H 240, MP 180, CB 60
  (b) Angles 165°, 75°, 60°, 45°, 15°
  (c) Angles 132°, 60°, 96°, 36°, 36°;
    11 : 5 : 8 : 3 : 3

3 (a) 5.52, 2.15
  (b) Median = 5.24, LQ = 3.8, UQ = 6.2
  (c) −0.2, 44%

4 (a) F 70°, A 88°, P 150°, W 52°
  (b) Ratios of profits constant.
    Profits ∝ Area, so increase.
    New radius = 4.29 cm

5 Angles   M 1°   SO 2°   MC 12°
  Numbers   4   8   52

  Angles   SB 29°   FW 266°   A 50°
  Numbers   128   1184   224

6 (a) Po 150° L 15° C 30° Pe 75° B 90°
  (b)   12   1   2   5   4
      180°   15°   30°   75°   60°
  (c) $R_3 : R_1 = \sqrt{3} : 1$; angles as in (a)

7 (a) Median = 5.15,
    IQR = 6.5 − 3.4 = 3.1
  (b) 69.7% ≈ 70%, 3.2 marks

*8 (a) One football pitch is represented
    by $\dfrac{164}{4} = 41°$

    One hockey pitch is represented
    by $\dfrac{54.7}{2} = 27.35°$

    So area of one hockey pitch =
    $\dfrac{27.35}{41} \times 0.753$ ha = 0.502 ha

  (b) Area of lacrosse pitch
    $= \dfrac{59.3}{41} \times 0.753 = 1.09$ ha

  (c) Total area $= \dfrac{360}{41} \times 0.753$
      = 6.61 ha

9 (a) 0.0502, 0.0133
  (b) 502, 133
  (c) 4.61, 1.28

10 (a) S 150, M 300, ST 600, L 450, EL 300
   (c) 55 g

11 (a) 46 minutes
   (b) Heights 0.6, 0.5, 1.2, 0.6, 0.4 units
   (c) (i) 6   (ii) 5

12 (a) Histogram heights 1.5, 2, 3.5, 2.5, 2, 1
   (b) (i) 3.40(48)   (ii) 3.86

13 (a) 289 approximately
   (c) Bar chart or line diagram

14 (a) 11, 9, 8, 1
   (b) Median = 14 min, LQ = 6 min, UQ = 23 min
   (c) 22 min
   (e) Small positive (quartile coefficient $= \frac{1}{17}$)

Revision

1 S 148°, V 48°, B 108°, E 56°; £25 000

2 60°, 100°, 140°, 60° for both years
  First radius = 4 cm

4 2 | 1 1 2 2 2 2
    2 | 3 4 4 4 4
    2 | 5 5 5 5 5 6 6 6
    2 | 7 7 7 7 7 8 8 8 8 8
    2 | 9 9
  Key: 2 | 4 means 24
  Class intervals: 20–21, ... , 30–31

5 20 | 5 6 8
    21 | 0 4 8
    22 |
    23 | 0 1 1 2 4 5 6 8
    24 | 1 3 3 5 6 7 7 8
    25 | 0 2 7
    26 | 4 5 7 8 9
  Key: 20 | 5 means 20.5

6 Computer 240°, £1000; printer 72°, £300; furniture and books 42°, £175; help 6°, £25

7

		Physics							Maths						
					8	2									
	9	7	3	3	1	0	3	3	4	8					
8	7	5	5	4	3	1	4	0	2	2	5	7	9	9	
	9	6	5	2	2	5	3	5	5	6	8	8	9		
		9	4	6	1	2	2	6							
			7	3	7	2	3	6							
			5	2	8	4									

Key: | 5 | 3 means 53
              2 | 5 | means 52
Maths: $\mu = 54.76$, $\sigma = 12.96$
Physics: $\mu = 50.8$, $\sigma = 16.25$

8

27 | 1 4 6 7 8 9
28 | 0 1 1 3 4 5 6 6 7 7 8 8 8
29 | 0 0 2 5 5
30 | 2

Key: 27 | 4 means 274
Note: This is just one suggestion for the answer.

9

		Females			Males							
			9	7								
			7	1	8							
		9	9	8	9							
9	9	8	5	5	3	1 1	10					
		8	6	5	5	2 0	11					
					12							
					13	5						
					14	1 3 8						
					15	6 7 9						
					16	0 2 3 4 5 7 8 8 9					
					17	0 2 3 5						

Key: 8 | 9 | means 980 g
              15 | 6   means 1560 g
mean (M) − mean (F) = 1607.5 − 1035
                         = 572.5 g

10

	UK		Abroad
28 28 28 28 23 22	1	1	00 11 22 25 28 39
22 12 11 11 00 00	1	39 50 61 72 72	
8	2	22 22 39	
	3	00 11	
	4		
	5		
	6		
8 8 2 2	7		
9 9 9	8		

Key: 8 | 7 | means 7.8 °C
             22 | 1 | means 12.2 °C
                1 | 00 means 10.0 °C
                3 | 11 means 31.1 °C
                6 | 1 means 6.1 °C

11 Histogram heights 6, 34, 46, 10, 4

*12 (a) Mid-range masses
              117.5   130   140   150   165

Frequency
          2   13   22   9   4

Mean estimate

$$= \tfrac{1}{50}(117.5 \times 2 + 130 \times 13 + 140$$
$$\times 22 + 150 \times 9 + 165 \times 4)$$
$$= \tfrac{7015}{50} = 140.5 \text{ kg}$$

(b) Heights of rectangles
    110–125   125–135   135–145
    $\frac{2}{15} = 0.13$ $\frac{13}{10} = 1.3$ $\frac{22}{10} = 2.2$

                   145–155   155–175
                   $\frac{9}{10} = 0.9$ $\frac{4}{20} = 0.2$

Chapter 2 Probability

2.1

Basic

1 (a) $\frac{1}{2}$ (b) $\frac{1}{3}$ (c) $\frac{1}{3}$

2 (a) $\frac{7}{10}$ (b) $\frac{1}{10}$ (c) $\frac{2}{5}$ (d) $\frac{3}{10}$

3 (a) $\frac{1}{4}$ (b) $\frac{1}{2}$ (c) $\frac{3}{4}$

4 (a) $\frac{8}{23}$ (b) $\frac{17}{23}$ (c) $\frac{9}{23}$

*5 $x + 2x + 3x = 12$, i.e. $x = 2$; red = 2, white = 4, blue = 6
  (a) p(red) $= \frac{2}{12} = \frac{1}{6}$
  (b) p(not white) $= \frac{8}{12} = \frac{2}{3}$
  (c) p(white or blue) $= \frac{10}{12} = \frac{5}{6}$

6 (a) $\frac{1}{4}$ (b) $\frac{1}{2}$ (c) $\frac{1}{2}$ (d) $\frac{1}{4}$
  (e) 0 (f) $\frac{3}{4}$

7 (a) $\frac{1}{13}$ (b) $\frac{1}{26}$ (c) $\frac{2}{13}$ (d) $\frac{11}{13}$

8 (a) $\frac{1}{4}$ (b) $\frac{1}{7}$ (c) $\frac{1}{3}$

9 (a) $\frac{1}{52}$ (b) $\frac{2}{51}$ (c) $\frac{39}{50}$

10 (a) $\frac{1}{4}$ (b) $\frac{3}{4}$

11 (a) $\frac{5}{18}$ (b) $\frac{35}{36}$

12 (b) (i) $\frac{21}{50}$ (ii) $\frac{51}{100}$

*13 (a) (i) p(even) $= \frac{10}{20} = \frac{1}{2}$
    (ii) p(prime) $= \frac{8}{20} = \frac{2}{5}$
    (iii) p(perfect square) $= \frac{4}{20} = \frac{1}{5}$
  (b) (i) p(add up to three)
    $= \frac{1}{20} \times \frac{1}{19} \times 2 = \frac{1}{190}$

(ii) p(perfect squares) $= \frac{4}{20} \times \frac{3}{19}$
               $= \frac{3}{95}$
(iii) p(both > 15) $= \frac{5}{20} \times \frac{4}{19} = \frac{1}{19}$.

Intermediate

1 (a) $\frac{5}{18}$ (b) $\frac{1}{2}$ (c) $\frac{5}{12}$ (d) $\frac{2}{3}$

2 (a) $\frac{1}{6}$ (b) $\frac{1}{2}$ (c) $\frac{1}{6}$ (d) $\frac{2}{9}$

3 (a) $\frac{2}{9}$ (b) $\frac{1}{9}$ (c) $\frac{2}{9}$

4 (a) $\frac{1}{78}$ (b) $\frac{1}{156}$ (c) $\frac{15}{26}$

*5 Winning numbers 529, 576, 625, 676, 729, 784, 841, 900, 961
  (a) p(first wins) = 9/1000
  (b) p(second fails) $= 1 - \frac{8}{999} = \frac{991}{999}$
  (c) p(3rd wins) $= \frac{8}{998} = \frac{4}{499}$
  (d) p(1st fails and 2nd wins)
    $= \frac{991}{1000} \times \frac{9}{999} = \frac{991}{111\,000}$

6 (a) $\frac{8}{15}$ (b) $\frac{2}{15}$ (c) $\frac{4}{15}$ (d) $\frac{2}{3}$

7 (a) $\frac{3}{8}$ (b) $\frac{3}{4}$ (c) $\frac{1}{8}$ (d) $\frac{1}{2}$

8 (a) $\frac{1}{16}$ (b) $\frac{1}{4}$ (c) $\frac{15}{16}$ (d) $\frac{5}{8}$

9 (a) $\frac{1}{10}$ (b) $\frac{9}{20}$ (c) $\frac{9}{20}$

10 (a) 0.63   (b) 0.216

11 (a) $\frac{1}{28}$ (b) $\frac{11}{14}$ (c) $\frac{17}{56}$

12 (a) $\frac{1}{8}$ (b) $\frac{1}{2}$ (c) $\frac{1}{2}$

13 (a) $\frac{1}{3}$ (b) $\frac{349}{599}$ (c) $\frac{125}{299}$

14 (a) $\frac{1}{15}$ (b) $\frac{8}{15}$ (c) $\frac{4}{15}$

15 (a) $\frac{4}{9}$ (b) $\frac{4}{9}$

16 (a) $\frac{25}{48}$ (b) $\frac{23}{48}$ (c) $\frac{1}{16}$ (d) $\frac{7}{12}$
    (e) $\frac{2}{3}$ (f) $\frac{1}{15}$ (g) $\frac{1}{12}$ (h) $\frac{1}{25}$

Advanced

1 (a) $\frac{1}{5}$ (b) $\frac{2}{5}$ (c) $\frac{11}{20}$ (d) $\frac{1}{120}$ (e) $\frac{1}{30}$

2 (a) $\frac{7}{18}$ (b) $\frac{5}{36}$ (c) $\frac{1}{4}$ (d) $\frac{1}{6}$

*3

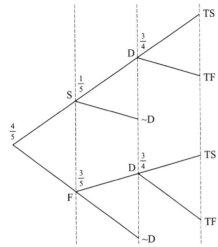

Key: S immunisation successful,
    F immunisation fails,
    D catches disease,
    ~D does not catch disease,
    TS treatment successful,
    TF treatment fails.

(a) $\mathrm{p}(\mathrm{F}, \sim \mathrm{D}) = \frac{1}{5} \times \frac{2}{5} = \frac{2}{25}$
(b) $\mathrm{p}(\mathrm{S}, \mathrm{D}, \mathrm{TF}) = \frac{4}{5} \times \frac{1}{5} \times \frac{1}{4} = \frac{1}{25}$
(c) p(treatment needed)
    $= \frac{4}{5} \times \frac{1}{5} + \frac{1}{5} \times \frac{3}{5} = \frac{7}{25}$

4 (a) $\frac{4}{9}$ (b) $\frac{3}{4}$ (c) $\frac{4}{9}$ (d) $\frac{5}{9}$ (e) $\frac{1}{9}$

5 (a) 0.18   (b) 0.08   (c) 0.5

6 (a) $\frac{12}{25}$ (b) $\frac{4}{25}$ (c) $\frac{7}{25}$

7 (a) (i) $\dfrac{R_3^2 - R_2^2}{R_3^2}$

    (ii) $\dfrac{R_2^2 - R_1^2}{R_3^2}$

    (iii) $\dfrac{R_1^2 - R_2^2 + R_3^2}{R_3^2}$

    (b) $\dfrac{4}{R_3^4}\{(R_2^2 - R_1^2)(R_1^2 - R_2^2 + R_3^2)\}$

8 (a) p^2 (b) $2p(1-p)$ (c) $\frac{2}{3}(1-p)^2$

9 (a) $\frac{4}{7}$ (b) $\frac{71}{84}$

10 (a) $\frac{1}{4}$

Revision

1 (a) $\frac{4}{27}$ (b) $\frac{20}{81}$ (c) $\frac{4}{81}$ (d) 0   (e) $\frac{2}{9}$

2 (a) $\frac{1}{26}$ (b) $\frac{1}{1820}$ (c) $\frac{1}{1820}$ (d) $\frac{9}{455}$

3 (a) $\frac{3}{100}$ (b) $\frac{9}{200}$ (c) $\frac{27}{200}$ (d) $\frac{7}{100}$

4 (a) (i) $\frac{1}{6}$ (ii) $\frac{1}{8}$ (iii) $\frac{13}{24}$
    (b) (i) $\frac{1}{16}$ (ii) $\frac{1}{32}$ (iii) $\frac{1}{12}$ (iv) $\frac{23}{144}$

5 (a) $\frac{3}{4}$ (b) $\frac{3}{10}$ (c) $\frac{33}{100}$

6 (a) $\frac{1}{110}$ (b) $\frac{1}{880}$ (c) $\frac{1}{6160}$ (d) $\frac{1}{9240}$

7 (a) (i) $\frac{5}{12}$ (ii) $\frac{5}{12}$ (iii) $\frac{1}{2}$
    (b) (i) $\frac{25}{144}$ (ii) $\frac{1}{12}$ (iii) $\frac{5}{18}$

8 (a) $\frac{1}{5}$ (b) $\frac{1}{10}$ (c) $\frac{13}{40}$

9 (a) $\frac{1}{132}$ (b) $\frac{1}{720}$ (c) $\frac{1}{120}$

10 (a) $\frac{11}{40}$ (b) $\frac{1}{20}$

*11 (a) (i) p(6, heads) $= \frac{3}{5} \times \frac{3}{10} = \frac{9}{50}$
        (ii) p(3 or 4, tails)
            $= \frac{2}{25} \times 2 \times \frac{7}{10} = \frac{14}{125}$
    (b) p(11, tails)
        $= \frac{3}{5} \times \frac{2}{25} \times 2 \times \frac{7}{10} = \frac{42}{625} = 0.0672$

12 (a) $\frac{1}{3}$ (b) (i) $\frac{1}{9}$ (ii) $\frac{4}{9}$
    (c) (i) $\frac{1}{1296}$ (ii) $\frac{8}{27}$

13 (a) $\frac{1}{57}$ (b) $\frac{21}{38}$ (c) $\frac{9}{38}$

14 Four darts

2.2

Basic

1 (a) 0.2   (b) $\frac{2}{3}$ (c) $\frac{1}{2}$ (d) 0.6
(e) 0.2

2 (a) 11   (b) 4   (c) $\frac{2}{7}$ (d) $\frac{5}{8}$ (e) 3

3 (a) $w = 9$, $x = 1$, $y = 4$, $z = 6$
(b) (i) $\frac{1}{2}$ (ii) $\frac{9}{20}$ (iii) $\frac{2}{5}$

4 (a) 0.3   (b) 0.5   (c) 0.6
(d) 0.2   (e) 0.8

*5 (a) p(same vowel twice)
$= \frac{2}{10} \times \frac{1}{10} = \frac{1}{50}$
(b) p(sum $= 4$) $= \frac{1}{10} \times \frac{1}{10} \times 2$
$+ \frac{1}{10} \times \frac{1}{10} = \frac{3}{100}$
(c) p(sum $= 10 \mid 2$ numbers drawn)
$= \frac{1}{5} \times \frac{1}{5} = \frac{1}{25}$

6 (a) (i) $\frac{1}{16}$ (ii) $\frac{1}{8}$ (iii) $\frac{9}{16}$ (b) $\frac{1}{3}$

7 (a) (i) $\frac{1}{3}$ (ii) $\frac{2}{3}$ (b) $\frac{1}{2}$

8 (a) $\frac{1}{4}$ (b) $\frac{2}{3}$ (c) $\frac{3}{8}$ (d) $\frac{3}{8}$

9 336

10 56; 6

11 (a) 60   (b) 120

12 560

Intermediate

1 40 320; $\frac{1}{2}$

2 (a) 9   (b) 840   (c) 1974   (d) 336

3 (a) 744   (b) 114   (c) 30   (d) 280
(e) 60   (f) 4.62   (g) 4320   (h) 280

(i) 3600   (j) 3.75   (k) 19
(l) $x(x + 1)(x + 2)$ (m) $\dfrac{1}{x - 2}$
(n) $\dfrac{x + 3}{x + 2}$ (o) $\dfrac{1}{x + 2}$

4 180;   (a) 30   (b) 120   (c) 60

5 61 236

6 0.75

*7 (a) p(G_1, B_1 in any order)
$= \frac{1}{10} \times \frac{1}{9} \times 2 = \frac{1}{45}$
(b) p(G, G) $= \frac{5}{10} \times \frac{4}{9} = \frac{2}{9}$
(c) p(B $\mid G_5$) $= \frac{2}{9}$

8 (a) $\frac{1}{57}$ (b) $\frac{18}{95}$ (c) $\frac{91}{228}$ (d) $\frac{9}{38}$

9 (a) $\frac{1}{2}$ (b) $\frac{1}{6}$ (c) $\frac{4}{9}$

10 (a) (i) 0   (ii) $\frac{1}{52}$
(b) (i) $\frac{2}{663}$ (ii) $\frac{1}{338}$
(c) (i) $\frac{26}{51}$ (ii) $\frac{1}{2}$
(d) (i) $\frac{4}{51}$ (ii) $\frac{1}{13}$
(e) (i) $\frac{1}{221}$ (ii) $\frac{1}{169}$

11 (a) $\frac{1}{77}$ (b) $\frac{1}{462}$ (c) $\frac{10}{77}$ (d) $\frac{25}{154}$

*12 (a) p(boy) $= \frac{3}{7}$
(b) (i) p(G, G) $= \frac{16}{28} \times \frac{15}{27} = \frac{20}{63}$
(ii) p(B, G in any order)
$= \frac{12}{28} \times \frac{16}{27} \times 2 = \frac{32}{63}$
(c) (i) p(B, G, B) $= \frac{12}{28} \times \frac{16}{28} \times \frac{12}{28} = \frac{36}{343}$
(ii) p(3G) $= \left(\frac{16}{28}\right)^3 = \frac{64}{343}$
(iii) p(more G than B)
$= \frac{144}{343} + \frac{64}{343} = \frac{208}{343}$

13 (a) $\frac{5}{8}$
(b) (i) $\frac{3}{28}$ (ii) $\frac{3}{14}$
(c) (i) $\frac{1}{20}$ (ii) $\frac{3}{5}$

14 (a) (i) $\frac{1}{9}$ (ii) $\frac{1}{20}$ (b) $\frac{9}{20}$

15 (a) $\frac{209}{600} = 0.348$ (b) $\frac{2}{3}$ (c) $\frac{7}{24}$

Advanced

1 81

2 (a) $\frac{13}{25}$ (b) $\frac{3}{10}$ (c) $\frac{4}{13}$ (d) $\frac{9}{25}$

3 (a) 0.12   (b) 0.7   (c) 0.15

*4 (a) p(germinate) = (0.4×0.45)
$+ (0.4 \times 0.6) + (0.2 \times 0.35) = 0.49$
(b) p($\sim$germinate | F_3) = $1 - 0.35 = 0.65$
(c) p(F_2 | $\sim$germinate)

$$= \frac{0.4 \times 0.4}{(0.4 \times 0.55) + (0.4 \times 0.4) + (0.2 \times 0.65)}$$

$$= 0.314$$

5 (a) (i) $\frac{1}{45}$ (ii) $\frac{14}{45}$ (iii) $\frac{2}{15}$
(b) (i) $\frac{5}{8}$ (ii) $\frac{3}{4}$

6 (a) 0.18   (b) 0.55   (c) 0.22

7 (a) (i) $\frac{1}{3003}$ (ii) $\frac{4}{3003}$ (iii) $\frac{1}{15}$
(b) $^{15}C_{11} = 1365$

8 (a) (i) 0.45   (ii) 0.25   (iii) 0.35
(b) (i) $\frac{7}{9}$ (ii) $\frac{2}{5}$ (iii) $\frac{1}{3}$

9 (i) 0.12   (ii) 0.72   (iii) $\frac{2}{3}$

10 (i) 65 780   (ii) 1.52×10^{-5}
(iii) $^5C_4(21) = 105$ (iv) 2100

Revision

1 9.23×10^{-6}

2 (a) 5040   (b) 50 400   (c) 2520
(d) 75 600   (e) 420

3 0.00139

4 (a) 6   (b) 48

5 $n = 6$

6 (a) (i) $\frac{2}{5}$ (ii) 1
(b) (i) (A) $\frac{33}{95}$ (B) $\frac{9}{25}$ (ii) (A) $\frac{48}{95}$ (B) $\frac{12}{25}$
(iii) (A) $\frac{8}{19}$ (B) $\frac{2}{5}$

7 (a) $\frac{1}{10}$ (b) $\frac{13}{20}$ (c) $\frac{3}{10}$ (d) $\frac{5}{11}$

8 $\frac{1}{60}$ and $\frac{1}{420}$

*9 p(3 aces, 2 kings)

$$= \frac{4}{52} \times \frac{3}{51} \times \frac{2}{50} \times \frac{4}{49} \times \frac{3}{48} \times 10$$

$$= \frac{2880}{52 \times 51 \times 50 \times 49 \times 48}$$

p(A, K, Q, J, 10 spades)

$$= \frac{1}{52} \times \frac{1}{51} \times \frac{1}{50} \times \frac{1}{49} \times \frac{1}{48} \times 5!$$

$$= \frac{120}{52 \times 51 \times 50 \times 49 \times 48}$$

So,

$$\frac{\text{p(3 aces, 2 kings)}}{\text{p(A, K, Q, J, 10 spades)}} = \frac{2880}{120} = 24$$

10 (a) $\frac{5}{28}$ (b) $\frac{1}{56}$ (c) $\frac{15}{28}$ (d) $\frac{5}{7}$

11 (a) 2   (b) 9; 161 280, 725 760

12 (a) (i) 180   (ii) 30   (iii) 453 600
(iv) 30 240   (v) 10 080
(b) 10

Chapter 3 Discrete distributions

3.1

Basic

1 (a) $\frac{1}{18}$ (b) 1   (c) $\frac{1}{8}$ (d) 3
  (e) $\frac{24}{13}$ (f) $\frac{1}{8}$ (g) $\frac{4}{7}$ (h) 32

3 (a) $\frac{1}{10}(x+1)$ (b) $\frac{1}{24}(2x-1)$
  (c) $\frac{1}{22}(x-6)$ (d) $\frac{1}{30}(x-1)^2$
  (e) $10x$

4

x	0	1	2	3
$p(X=x)$	$\frac{8}{125}$	$\frac{36}{125}$	$\frac{54}{125}$	$\frac{27}{125}$

5

x	0	1	2	3
$p(X=x)$	$\frac{8}{27}$	$\frac{12}{27}$	$\frac{6}{27}$	$\frac{1}{27}$

6

x	0	1	2
$p(X=x)$	$\frac{25}{36}$	$\frac{10}{36}$	$\frac{1}{36}$

7 (a) 3, 10   (b) 6.85, 47.7
  (c) -2, 5.2   (d) 3.4, 12.6
  (e) 11.7, 136   (f) 1.66, 3.86
  (g) 26.7, 714   (h) 0.02, 1.66
  (i) 4.88, 25.1   (j) 22.2, 496

8 (a) (i) 3.1   (ii) 12.4   (iii) 9.4
  (iv) 10.1   (v) 19.2
  (b) (i) 0.27   (ii) 0.135   (iii) 0.511
  (iv) 2.51   (v) 0.081
  (c) (i) 3.1   (ii) 31   (iii) 26   (iv) 22.7
  (d) (i) -1 (ii) -2 (iii) 7   (iv) 3
  (e) (i) -2.4 (ii) 13.2   (iii) -19.8

9 (a) 1   (b) 1.5

* 10 (a) (i) 2.75   (ii) 8.25   (iii) 0.688
  (iv) 2.75

*(b) (i) $E(X) = 2 \times \frac{1}{4} + 3 \times \frac{3}{8} + 4 \times \frac{1}{4}$
        $+ 5 \times \frac{1}{8} = \frac{13}{4} = 3\frac{1}{4}$
  (ii) $E(X^2) = 4 \times \frac{1}{4} + 9 \times \frac{3}{8} + 16 \times \frac{1}{4}$
        $+ 25 \times \frac{1}{8} = \frac{23}{2} = 11\frac{1}{2}$
  (iii) $Var(X) = 11\frac{1}{2} - (3\frac{1}{4})^2 = \frac{15}{16}$
  (iv) $Var(3X-1) = 9 \times \frac{15}{16} = 8\frac{7}{16}$
  (c) (i) -0.9 (ii) 5   (iii) 1.19
  (iv) 1.04
  (d) (i) 0.27   (ii) 0.249   (iii) 0.0404
  (iv) 0.100
  (e) (i) 2.05   (ii) 8.3   (iii) 0.362
  (iv) 1.20

11 $a = 3$

12 $b = 6$

13 (a) $\frac{1}{2}, \frac{5}{12}$ (b) 0.2, 0.18

Intermediate

1 (a) $\frac{1}{4}, \frac{31}{144}$

x	0	1	2
$p(X=x)$	$\frac{55}{72}$	$\frac{16}{72}$	$\frac{1}{72}$

  (b) (i) 1.2, 0.36

x	0	1	2
$p(X=x)$	$\frac{2}{20}$	$\frac{12}{20}$	$\frac{6}{20}$

  (ii) 1.8, 0.36

x	0	1	2	3
$p(X=x)$	0	$\frac{6}{20}$	$\frac{12}{20}$	$\frac{2}{20}$

  (c) 0.5, 0.33

x	0	1	2
$p(X=x)$	0.54	0.42	0.04

2 $d = 3$

3 (a)

x	0	1	2	3
$p(X=x)$	$\frac{27}{64}$	$\frac{27}{64}$	$\frac{9}{64}$	$\frac{1}{64}$

  (b) $\frac{27}{8000}$

4 (a)

x	0	1	2
$p(X = x)$	$\frac{5}{9}$	$\frac{2}{9}$	$\frac{2}{9}$

(b) $E(X) = \frac{2}{3}$

5 Red:

x	0	1	2
$p(X = x)$	$\frac{9}{48}$	$\frac{6}{48}$	$\frac{1}{48}$

Yellow:

x	0	1	2
$p(X = x)$	$\frac{50}{192}$	$\frac{60}{192}$	$\frac{18}{192}$

6 (a)

x	0	1	2
$p(X = x)$	$\frac{144}{169}$	$\frac{24}{169}$	$\frac{1}{169}$

(b)

x	0	1	2
$p(X = x)$	$\frac{9}{16}$	$\frac{6}{16}$	$\frac{1}{16}$

(c)

x	0	1	2
$p(X = x)$	$\frac{100}{169}$	$\frac{60}{169}$	$\frac{9}{169}$

7 (a) $E(X) = 1.8$, $\sigma^2 = 0.56$ (b) 2.11

8 (a) 3, 1 (b) 6.5, 1.75 (c) 2.19, 1.77
(d) 4.5, 1.75 (e) 0, 3.4

9 2.5

*10 (a) $k(2 + 3 + 4 + 5) + 2k(5 + 6 + 7)$
$= k(14 + 36) = 1$. So $k = \frac{1}{50}$
(b) $E(X)$
$= k(1 \times 2 + 2 \times 3 + 3 \times 4$
$+ 4 \times 5) + 2k(25 + 36 + 49)$
$= \frac{260}{50} = 5.2$
(c) $E(X^2)$
$= \frac{1}{50}(1 \times 2 + 4 \times 3 + 9 \times 4$
$+ 16 \times 5 + 250 + 432 + 686)$
$= \frac{1498}{50}$
So $\text{Var}(X) = \frac{1498}{50} - 5.2^2 = 2.92$

11 (a)

x	0	1	2	3
$p(X = x)$	$\frac{8}{27}$	$\frac{12}{27}$	$\frac{6}{27}$	$\frac{1}{27}$

(b) (i) 1   (ii) $\frac{2}{3}$
(c)

y	0	1	2	3
$p(Y = y)$	$\frac{1}{8}$	$\frac{3}{8}$	$\frac{3}{8}$	$\frac{1}{8}$

$E(Y) = \frac{3}{2}$, $\text{Var}(Y) = \frac{3}{4}$

z	0	1	2	3
$p(Z = z)$	$\frac{125}{216}$	$\frac{75}{216}$	$\frac{15}{216}$	$\frac{1}{216}$

$E(Z) = \frac{1}{2}$, $\text{Var}(Z) = \frac{5}{12}$

*12 (a)

x	0	1	2
$p(X = x)$	$\frac{81}{100}$	$\frac{18}{100}$	$\frac{1}{100}$

(i) $\frac{1}{5}$ (ii) $\frac{9}{50}$

*(b)

y	2	3	4	5
$p(Y = y)$	$\frac{1}{36}$	$\frac{2}{36}$	$\frac{3}{36}$	$\frac{4}{36}$

6	7	8	9	10	11	12
$\frac{5}{46}$	$\frac{6}{36}$	$\frac{5}{36}$	$\frac{4}{36}$	$\frac{3}{36}$	$\frac{2}{36}$	$\frac{1}{36}$

(i) $E(Y) = \frac{1}{36}(2 \times 1 + 3 \times 2 + 4$
$\times 3 + \cdots + 11 \times 2$
$+ 12 \times 1) = \frac{252}{36} = 7$
(ii) $E(Y^2) = \frac{1}{36}(4 \times 1 + 9 \times 2 + 16$
$\times 3 + \cdots + 121 \times 2$
$+ 144 \times 1) = \frac{1974}{36}$
So $\text{Var}(Y) = \frac{1974}{36} - 49 = 5.83$

13 (a) 4.32   (b) 61.9   (c) $8.64 + A$
(d) 7.96   (e) 39.6

14 (a)

x	1	2	3	4	5
$p(X = x)$	$\frac{5}{15}$	$\frac{4}{15}$	$\frac{3}{15}$	$\frac{2}{15}$	$\frac{1}{15}$

(b) (i) $\frac{7}{3}$ (ii) 1.56

15 $x = 2$

Advanced

1 (a) $\frac{5}{12}$ (b)

x	0	1	2
$p(X = x)$	$\frac{4}{9}$	$\frac{4}{9}$	$\frac{1}{9}$

(c) (i) $\frac{2}{3}$ (ii) $\frac{4}{9}$ (d) 1

*2 (a)

x	1	2	3	4
$p(X = x)$	$\frac{1}{8}$	$\frac{1}{8}$	$\frac{3}{8}$	$\frac{3}{8}$

(b) (i) $E(X) = \frac{1}{8}(1 \times 1 + 2 \times 1 + 3$
$\times 3 + 4 \times 3) = \frac{24}{8}$
$= 3$
(ii) $E(X^2) = \frac{1}{8}(1 \times 1 + 4 \times 1 + 9$
$\times 3 + 16 \times 3)$
$= 10,$
so $\text{Var}(X) = 10 - 3^2 = 1$

3 (a) $\frac{47}{100}$
(b)

x	0	1	2	3
$p(X = x)$	$\frac{6}{100}$	$\frac{35}{100}$	$\frac{47}{100}$	$\frac{12}{100}$

(c) (i) 1.65 (ii) 0.588

4 (a)

d	0	1	2	3	4
$p(D=d)$	$\frac{11}{230}$	$\frac{143}{575}$	$\frac{234}{575}$	$\frac{143}{575}$	$\frac{11}{230}$

5 (a) $\frac{1}{94}$

(b)

x	0	2	4	6	8	10	12
$p(X=x)$	$\frac{2}{94}$	$\frac{6}{94}$	$\frac{18}{94}$	$\frac{38}{94}$	$\frac{8}{94}$	$\frac{10}{94}$	$\frac{12}{94}$

(c) (i) 6.60   (ii) 13.2   (iii) 35.3

6 (a) 3   (b) 10.4   (c) 260

7 (a) 1.78   (b) 1.18

8 (a) (i) 1.94   (ii) 1.04
(b) 5 weeks, 70 days

9 (i) 30k, 12k, 2k
(ii) Modal value = 1, positively skewed
(iii) E(X) = 2, SD = 1.10   (iv) £1.70

Revision

1 (a) $\sum p(X=x) = \frac{6}{5} \Rightarrow$ no
(b) $\sum p(X=x) = \frac{53}{52} \Rightarrow$ no
(c) $\sum p(X=x) = 1 \Rightarrow$ yes
(d) $\sum p(X=x) = \frac{26}{27} \Rightarrow$ no
(e) $\sum p(X=x) = 1 \Rightarrow$ yes

2 (a)

x	0	1	2
$p(X=x)$	$\frac{3}{14}$	$\frac{8}{14}$	$\frac{3}{14}$

(b)

x	0	1	2
$p(X=x)$	$\frac{1}{4}$	$\frac{1}{2}$	$\frac{1}{4}$

3 A = 2.4

*4 (a) (i) $E(M) = \frac{1}{10}(0 \times 1 + 1 \times 4 + 2 \times 2 + 3 \times 2 + 4 \times 1)$
$= 1.8$
(ii) $E(M^2) = \frac{1}{10}(0 \times 1 + 1 \times 4 + 4 \times 2 + 9 \times 2 + 16 \times 1)$
$= 4.6$,
so $Var(X) = 4.6 - 1.8^2 = 1.36$
(b) E(no. of mistake-free performances) $= \frac{1}{10} \times 38 = 3.8 \approx 4$

5

x	0	1	2	3	4	5	6
$p(X=x)$	$\frac{1}{12}$	$\frac{2}{12}$	$\frac{2}{12}$	$\frac{2}{12}$	$\frac{2}{12}$	$\frac{2}{12}$	$\frac{1}{12}$

(a) 3   (b) 3.17   (c) $\frac{5}{12}$ (d) 60

6 (a)

g	0	1	2	3
$p(G=g)$	$\frac{5}{120}$	$\frac{23}{120}$	$\frac{54}{120}$	$\frac{38}{120}$

(b) (i) 2.04   (ii) 0.821

7 (a)

Score s	1	2	3	4
$p(S=s)$	$\frac{36}{200}$	$\frac{44}{200}$	$\frac{48}{200}$	$\frac{72}{200}$

(b) 1.25

8 (a)

x	1	2	3	4
$p(X=x)$	$\frac{9}{12}$	$\frac{9}{44}$	$\frac{9}{220}$	$\frac{1}{220}$

(b) 1.30, 0.319

9 $E(X) = 2.02$

10 (a)

n	0	1	2
$p(N=n)$	$\frac{25}{36}$	$\frac{10}{36}$	$\frac{1}{36}$

$E(N) = \frac{1}{3}$

(b)

k	0	1	2
$p(K=k)$	$\frac{16}{36}$	$\frac{16}{36}$	$\frac{4}{36}$

$E(K) = \frac{2}{3}$

11 (a) 3.75, 5.4, 5   (b) 4.44, 12.2, 11

*12 Total area $= \frac{1}{2}(8 + 14)5 = 55$. Areas for scores 2, 4 and 8 are 20, 20 and 15 resp. So, $p(2) = \frac{20}{55}$, $p(4) = \frac{20}{55}$, $p(8) = \frac{15}{55}$, with one dart.
With two darts,
$p(4) = \frac{20}{55} \times \frac{20}{55} = \frac{16}{121}$,
$p(6) = \frac{20}{55} \times \frac{20}{55} \times 2 = \frac{32}{121}, \ldots,$
$p(16) = \frac{15}{55} \times \frac{15}{55} = \frac{9}{121}$
The pdf is:

s	4	6	8	10	12	16
$p(S=s)$	$\frac{16}{121}$	$\frac{32}{121}$	$\frac{16}{121}$	$\frac{24}{121}$	$\frac{24}{121}$	$\frac{9}{121}$

3.2

Basic

1 (a) (i) 0.000 977   (ii) 0.237
(iii) 0.0879   (iv) $\frac{5}{4}$ (v) $\frac{15}{16}$
(b) (i) 0.00391   (ii) 0.00391
(iii) 0.219   (iv) 4   (v) 2
(c) (i) 0.263   (ii) 0.649   (iii) 0.00165
(iv) 2   (v) 1.15

(d) (i) 0.318 (ii) 0.0972
(iii) 0.00379 (iv) 2.1 (v) 1.47
(e) (i) 0.410 (ii) 0.263 (iii) 0.9997
(iv) 1 (v) 0.894
(f) (i) 0.0531   (ii) 0.0464
(iii) 0.251 (iv) 6 (v) 1.55
(g) (i) 0.484 (ii) 0.133 (iii) 0.0128
(iv) 1.75 (v) 1.31
(h) (i) 0.633 (ii) 0.104 (iii) 0.633
(iv) 3.75 (v) 0.968
(i) (i) 0.547 (ii) 0.0938
(iii) 3 (iv) 1.5
(j) (i) k^5 (ii) $5(k^4 - k^5)$
(iii) $5k(1 - k)^4$ (iv) $5k$
(v) $5k(1 - k)$
(k) (i) $Np(1 - p)^{N-1}$
(ii) $N(p^{N-1} - p^N)$
(iii) Np
(iv) $Np(1 - p)$
(l) (i) $Np(1 - \frac{1}{2}p)^{2N-1}$
(ii) $2N(\frac{1}{2}p)^{2N-1}(1 - \frac{1}{2}p)$
(iii) Np
(m) (i) $10p^2(1 - p)^{10p-1}$
(ii) $10p^{10p}(1 - p)$
(iii) $10(p^2 - p^3)$

2 (a) 0.146   (b) 0.244   (c) 0.0777
(d) 2.5   (e) 1.875

3 (a) 0.256   (b) 0.218; 300

8 (a) 0.216   (b) 0.288   (c) 0.432

9 (a) 0.817   (b) 0.0153   (c) 0.984

10 (a) 0.377   (b) 0.0415   (c) 0.776

11 (a) 7.29×10^{-7} (b) 0.817 (c) 0.167

Intermediate

1 (a) 0.376   (b) 0.357   (c) 0.242

2 (a) (i) 0.165 (ii) 0.996 (iii) 0.210
(b) 33

3 (a) (i) 0.00309 (ii) 0.0746 (iii) 0.117
(b) 80

4 $B(\frac{3}{4} + \frac{1}{4})^7$
(a) p(fewer than 2 rainy days)
   $= (\frac{3}{4})^7 + 7(\frac{3}{4})^6(\frac{1}{4}) = 0.445$
(b) p(fine for 5 + days)
   $= (\frac{3}{4})^7 + 7(\frac{3}{4})^6(\frac{1}{4}) + {}^7C_2(\frac{3}{4})^5(\frac{1}{4})^2$
   $= 0.756$
(c) p(only last 2 days fine)
   $= (\frac{1}{4})^5(\frac{3}{4})^2 = 0.000549$
(d) p(4 consecutive days of rain)
   $= (\frac{1}{4})^4(\frac{3}{4})^3 \times 4 = 0.00659$

4

x	0	1	2	3	4	5
(a) $p(X = x)$	0.327 68	0.4096	0.2048	0.0512	0.0064	0.000 32
(b)	$\frac{81}{256}$	$\frac{108}{256}$	$\frac{54}{256}$	$\frac{12}{256}$	$\frac{1}{256}$	—
(c)	0.0081	0.0756	0.2646	0.4116	0.2401	—
(d)	A^3	$3A^2(1 - A)$	$3A(1 - A)^2$	$(1 - A)^3$	—	—
(e)	$(1 - 2p)^3$	$6p(1 - 2p)^2$	$12p^2(1 - 2p)$	$8p^3$	—	—

5 (a)

x	0	1	2	3	4	5
$p(X = x)$	0.010 24	0.0768	0.2304	0.3456	0.2592	0.077 76

(b) (i) 0.2592 (ii) 0.01536

6 (a) p(no sixes) $= (\frac{5}{6})^2 = 0.335$
(b) p(two sixes) $= {}^6C_2(\frac{5}{6})^4(\frac{1}{6})^2 = 0.201$
(c) p(second and third rolls give a six)
   $= \frac{5}{6} \cdot \frac{1}{6} \cdot \frac{1}{6} \cdot \frac{5}{6} \cdot \frac{5}{6} \cdot \frac{5}{6} = 0.0134$

7 (a) $\frac{1}{64}$ (b) $\frac{9}{64}$ (c) $\frac{27}{64}$

5 (a) (i) 0.00415 (ii) 0.996 (b) 15

6 (a) 0.313 (b) 0.230

7 (a) 4 (b) 3

8 (a) (i) 0.0498 (ii) 0.106 (b) $a = 6$

9 (a) 0.346 (b) 0.0154 (c) 0.683
(d) 0.490 (e) 0.0691

10 (a) (i) 0.309 (ii) 0.00243 (iii) 0.0309
(b) (i) 0.313 (ii) 0.0313 (iii) 0.0313

11 $n_1 = 3$, $n_2 = 4$, $p_1 = \frac{1}{2}$, $p_2 = \frac{3}{4}$

*12 (a) $E(X_2) = 2p(n+1) = 2np + 2p$
$2\{E(X_1) + p\} = 2(np + p)$
$= 2np + 2p = E(X_2)$

(b) $\mathrm{Var}(X_2) = (n+1)2p(1-2p)$
$= (2np + 2p)(1 - 2p)$
$E(X_2)(1 - 2p) = (2np + 2p)$
$\times (1 - 2p)$
$= \mathrm{Var}(X_2)$

(c) $\mathrm{Var}(X_2) = (2np + 2p)(1 - 2p)$
$= 2np - 4np^2 + 2p - 4p^2$
$= 2np - 2np^2 - 2np^2$
$+ 2p - 4p^2$
$= 2np(1 - p)$
$+ 2p(1 - np - 2p)$
$= 2\,\mathrm{Var}(X_1)$
$+ 2p(1 - np - 2p)$

13 0.263

14 0.313, 0.00298

15 $n = 25$

Advanced

1 $(1 - p)^2 \Rightarrow$ TT, $2p(1 - p) \Rightarrow$ TH or
HT, $p^2 \Rightarrow$ HH; $3(p^2 - p^3)$)

2 (a) (i) 0.185 (ii) 0.0216 (iii) 0.0370
(b) $E(X) = 1.8$, $\mathrm{Var}(X) = 1.26$; 90

3 $E(X_1) = np$, $E(X_2) = 6np$,
$\mathrm{Var}(X_1) = np(1 - p)$,
$\mathrm{Var}(X_2) = 6np(1 - 3p)$

4

x	0	1	2	3
$p(X = x)$	0.216	0.432	0.288	0.064
No. of people	11	22	14	3

5 Mean $= 3$, Variance $= 0.76$, $p = \frac{3}{4}$

x	0	1	2	3	4
$p(X = x)$	$\frac{1}{256}$	$\frac{12}{256}$	$\frac{54}{256}$	$\frac{108}{256}$	$\frac{81}{256}$
Theoretical frequency	0.78125	9.375	42.1875	84.375	63.28125

*6 (a) $B(\frac{7}{8} + \frac{1}{8})^{20}$
(i) $p(5\ \mathrm{LH}) = {}^{20}C_5(\frac{7}{8})^{15}(\frac{1}{8})^5$
$= 0.0638$
(ii) $p(\text{more than } 17\ \mathrm{RH})$
$= (\frac{7}{8})^{20} + 20(\frac{7}{8})^{19}(\frac{1}{8})$
$+ {}^{20}C_2(\frac{7}{8})^{18}(\frac{1}{8})^2 = 0.535$

(b) $E(X) = 20(\frac{1}{8}) = 2.5$
(c) $p(3\ \mathrm{LH}) = {}^{20}C_3(\frac{7}{8})^{17}(\frac{1}{8})^3 \times 40 = 9$
to the nearest whole number

7 (a) (i) 0.117 (ii) 0.996
(iii) 3.66×10^{-5}
(b) 320

8 (a) 0.939 (b) 0.0435 (c) 0.0172

9 (a) (i) 0.00338 (ii) 0.0574
(iii) 0.168
(b) (i) 0.0327 (ii) 0.0923 (iii) 0.157

10 (a) 0.733 (b) 0.0703

11 (a) (i) $\frac{3}{8}$ (ii) $\frac{4}{15}$
(b) (i) $\frac{27}{125}$ (ii) $\frac{8}{125}$ (iii) $\frac{38}{125}$

Revision

1 (a) 0.086 783 8 (b) 0.112 275 1
(c) 0.202 808 8 (d) 0.115 054 2
(e) 0.092 158 9

2 (a) 0.0357   (b) 0.323; 2

3 $E(X) = 2.25$

4 (a) 0.0102   (b) 0.259

5 (a) 0.778   (b) 0.0241   (c) 0.999 880 2

6 (a) $p = 0.2$, $q = 0.8$, $n = 10$
  (b) (i) 0.302   (ii) 0.202

7 10

8 $n_1 = 10$, $n_2 = 9$, $p_1 = \frac{1}{4}$, $p_2 = \frac{2}{3}$

9 (a) 0.311   (b) 0.179
  (c) 0.0553   (d) 0.0311

10 5

11 (a) 0.0749
   (b) 2.25×10^{-5}
   (c) 0.00932

*12 (a) p(5 or 6 first) $= \frac{1}{3}$
   (b) $B(\frac{2}{3} + \frac{1}{3})^{50}$ p(5 or 6, five times)
     $= {}^{50}C_5(\frac{2}{3})^{45}(\frac{1}{3})^5 = 1.04 \times 10^{-4}$
     $= 0.000\ 104$
   (c) p(5 or 6, ten times)
     $= {}^{50}C_{10}(\frac{2}{3})^{40}(\frac{1}{3})^{10} = 0.0157$

3.3

Basic

1 (a) 0.24   (b) 0.0384   (c) 0.064
   (d) 0.0102   (e) 0.936   (f) 0.996
   (g) 0.064   (h) 0.16   (i) $\frac{5}{3}$
   (j) $\frac{10}{9}$ (k) 1

2 (a) 0.147   (b) 0.0720   (c) 0.657
   (d) 0.832   (e) 0.51   (f) 0.832
   (g) 0.49   (h) 0.240   (i) $\frac{10}{3}$
   (j) 2.79   (k) 0.240   (l) 0.49

3 (a) $\frac{11}{144}$ (b) 0.294

*4 Distribution is Geo(0.05).
   (a) p(catch on third cast)
     $= (0.95)^2(0.05) = 0.0451$
   (b) p(catch after 20th cast)
     $= (1 - 0.05)^{20} = (0.95)^{20}$
     $= 0.358$

5 (a) 0.0122   (b) 0.244

6 (a) (i) 0.224   (ii) 0.647
     (iii) 0.353   (iv) 3
   (b) (i) 0.180   (ii) 0.677
     (iii) 0.0902   (iv) 2
   (c) (i) 0.0668   (ii) 0.544
     (iii) 0.257   (iv) 1.58
   (d) (i) 0.195   (ii) 0.762
     (iii) 0.433   (iv) 2
   (e) (i) 0.270   (ii) 0.161
     (iii) 0.380   (iv) 1.45

7 (a) 0.101   (b) 0.917   (c) 0.459
   (d) 0.161   (e) 0.151   (f) 0.830
   (j) 0.0145   (h) 0.525

8 (a) 0.105   (b) 0.609

9 (a) $\frac{1}{4}$ (b) 0.779

*10 Distribution is Po(2) for 50 clips.
   (a) p(less than 5 faulty)
     $$= e^{-2}\left(1 + 2 + \frac{2^2}{2!} + \frac{2^3}{3!} + \frac{2^4}{4!}\right)$$
     $$= e^{-2} \times 7 = 0.947$$
   (b) p(none faulty) $= e^{-2} = 0.135$
   (c) p(6 or 7 or 8 faulty)
     $$= e^{-2}\left(\frac{2^6}{6!} + \frac{2^7}{7!} + \frac{2^8}{8!}\right) = 0.0163$$

11 (a) $\lambda = 7.5$ (i) 0.0156 (ii) 0.995
     (iii) 0.433
   (b) $\lambda = 4.8$ (i) 0.008 23 (ii) 0.0147
     (iii) 0.429   (iv) 0.000 104
   (c) $\lambda = 9$ (i) 0.001 23 (ii) 0.132
     (iii) 0.207

12 (a) 0.0183   (b) 0.0916   (c) 0.567

Intermediate

1 (a) 10   (b) 0.0656

2 (a) $\frac{2}{9}$ (b) 0.008 67   (c) 0.247
(d) 0.132

3 (a) $\frac{1}{3}$ (b) $\frac{8}{81}$ (c) $\frac{4}{9}$; 3

*4 (a) p(S wins on second throw)
$= (0.7)^2(0.5)^2 = 0.123$
(b) p(YL wins on second throw)
$= 0.7 \times 0.5 \times 0.3 = 0.105$
(c) p(YL wins)
$= 0.3 + (0.7 \times 0.5)(0.3)$
$+ (0.7 \times 0.5)^2(0.3) + \cdots$
= the sum to ∞ of a GP with
1st term 0.3 and
CR $(0.7 \times 0.5) = 0.35$
$= \dfrac{0.3}{1 - 0.35} = 0.462$

5 (a) 0.0819   (b) 0.002 75
(c) 0.001 89

6 (a) $p = \frac{1}{2}$ (b) $p = \frac{2}{3}$ (c) 0.006 14

7 $E(X) = 1$

8 (a) 0.570   (b) 0.0357

9 (a) (i) 0.167   (ii) 0.165
(b) (i) 0.828   (ii) 0.827

*10 (a) $\dfrac{e^{-\lambda} \cdot \lambda^2}{2!} + \dfrac{1}{4}e^{-\lambda} \cdot \dfrac{\lambda^3}{6} = \dfrac{e^{-\lambda} \cdot \lambda^4}{24}$,

so $\dfrac{1}{2} + \dfrac{\lambda^2}{24} = \dfrac{\lambda^2}{24}$

i.e. $\lambda^2 - \lambda - 12 = 0$
$\Rightarrow (\lambda - 4)(\lambda + 3) = 0$,
so $\lambda = 4$

(b) $p(X = 4) = \dfrac{e^{-4} \cdot 4^4}{4!} = 0.195$

11 (a) (i) 0.322   (ii) 0.453
(b) (i) 0.376   (ii) 0.408

12 (a) 0.06   (b) 0.0157   (c) 0.0698

13 (a) 0.147   (b) 0.550   (c) 0.0498

14 (a) 0.113   (b) 0.175   (c) 0.001 43

15 (a) 2.00   (b) 0.677   (c) 0.153

Advanced

*1 (a) (i) p(first three the same)
$= (\frac{1}{5})^2 = \frac{1}{25}$
(b) p(first four different)
$= 1 \times \frac{4}{5} \times \frac{3}{5} \times \frac{2}{5} = \frac{24}{125}$
(b) p(Crasher and Dasher)
$= \frac{1}{5} \times \frac{1}{5} \times 2 = \frac{2}{25}$
(c) $p(X > N) = (0.8)^N$
$\Rightarrow p(X < N) = 1 - (0.8)^N$
$\Rightarrow 1 - (0.8)^N > 0.8$,
so $N > \dfrac{\log 2}{\log 8} > 7.2$.
This gives $N_{\min} = 8$

2 (a) (i) $\frac{5}{36}$ (ii) $\frac{125}{1296} = 0.0965$
(b) (i) $\frac{25}{216}$ (ii) $\frac{125}{466\,56} = 0.002\,68$
(c) 3 or less

3 (a) 0.101   (b) 0.191   (c) 0.155
(d) 0.132

4 (a) 0.0166   (b) (i) 0.996 (ii) 0.995

5 (a) (i) 0.0378   (ii) 0.0948
(b) (i) 0.002 48   (ii) 0.0155   (iii) 0.161

6 (a) (i) 0.0613   (ii) 0.736
(b) (i) 0.123   (ii) 0.003 42   (iii) 0.134

7 (a) 0.0902   (b) 0.156   (c) 0.134
(d) 0.978   (e) 0.0529

8 (a) 7.5
(b) No. of raisins per bun constant
(d) 0.901

9 (a) 0.7787   (b) 0.1254

10 (a) 0.3782   (b) 0.3233   (c) 0.0043
  (d) (i) Not necessary – four clerks
           should cope
      (ii) No – family groups arrive
           together
     (iii) No – arrivals are not
           independent nor at constant
           rate.
      (iv) Tickets may have been
           bought earlier, group may
           have one ticket.

Revision

*1 (a) $p(X \leq 2) = (0.8)^0(0.2)$
$$+ (0.8)^1(0.2)$$
$$= 0.2 + 0.16 = 0.36$$
  (b) $p(X \leq 4) = 1 - (0.8)^4 = 0.590$
  (c) $p(X > 3) = (0.8)^3 = 0.512$
  (d) $p(X > 1) = (0.8)^1 = 0.8$
  (e) $p\{(X > 3)|(X > 1)\}$
$$= p(X > 2) = (0.8)^2 = 0.64$$
  (f) $p\{(X > 4)|(X > 1)\}$
$$= p(X > 3) = (0.8)^3 = 0.512$$

2 $k = \frac{1}{2}$

*3 For three matches, $X = Po(6)$.
  (a) $p(X < 2) = p(X = 0) + p(X = 1)$
$$= e^{-6}(1 + 6) = 0.0174$$

  (b) $p(X > 5)$
$$= 1 - p(X \leq 5)$$
$$= 1 - \{p(X = 0) + \cdots + p(X = 5)\}$$

$$= 1 - \left\{ e^{-6}\left(1 + 6 + \frac{6^2}{2!} + \frac{6^3}{3!} + \frac{6^4}{4!} + \frac{6^5}{5!}\right) \right\}$$

$$= 1 - 179.8e^{-6} = 0.554$$

$$p(8 \text{ goals}) = p(X = 8) = \frac{e^{-6} \times 6^8}{8!}$$

$$= 0.103 \approx 10\%$$

4 (a) 0.384   (b) 0.805   (c) 0.003 42

5 (a) 0.140   (b) 0.125   (c) 0.138
  (d) 0.151   (e) 0.119

6 (a) 0.0922   (b) 0.0188

7 (a) (i) 0.186   (ii) 0.184
  (b) (i) 0.215   (ii) 0.212
  (c) (i) 0.0332   (ii) 0.0337

8 (a) 0.006 74   (b) 0.104   (c) 0.384

9 (a) (i) $\frac{1}{16}$ (ii) $\frac{7}{16}$ (b) 0.0549

10 (a) (i) 0.125 (ii) 0.0948
   (b) (i) 3.04×10^{-5} (ii) 0.300

11 (a) 3   (b) 0.224

12 (a) (i) 3   (ii) 0.224
   (b) (i) 6   (ii) 0.161

Chapter 4 Continuous distributions

4.1

Basic

Answers for f(x) and F(x) are
abbreviated, see question 8(i).

1 (a)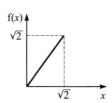

(b)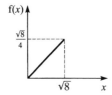

(c)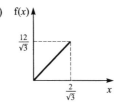

(d)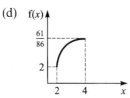

(e)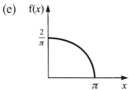

* 2 (a)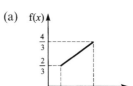

$k = \frac{1}{3}$

(b)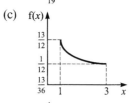

$k = \frac{6}{19}$

(c)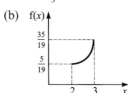

$k = \frac{1}{12}$

(d)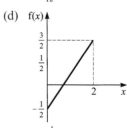

$k = \frac{1}{2}$

(e)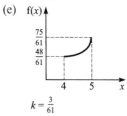

$k = \frac{3}{61}$

*(f)   f(x)

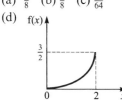

$$4k \int_2^5 (x^2 - 1)dx = 4k\{\tfrac{125}{3} - 5)$$
$$-(\tfrac{8}{3} - 2) = 1$$
$$\Rightarrow 4k \times 36 = 1,$$
so $k = \tfrac{1}{144}$

(g)   f(x)

1.8
$\tfrac{2}{15}$

           1   3   x

$k = \tfrac{9}{10}$

(h)   f(x)

$\tfrac{2}{3}$

      0   3   x

$k = \tfrac{\sqrt{2}}{3}$

3   (a) $\tfrac{1}{8}$ (b) $\tfrac{7}{8}$ (c) $\tfrac{7}{64}$

(d)   f(x)

$\tfrac{3}{2}$

      0   2   x

4   (a) $\tfrac{1}{12}$

(b)   f(x)   (c) $\tfrac{2}{3}$

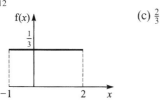

5   (a) Yes   (b) No   (c) No

$$\int_1^3 \tfrac{1}{4} x \, dx = 1 \qquad \int_{-2}^4 \tfrac{x^2}{8} dx = 3$$

$$\int_1^3 (4x^3 - 1)dx = 78$$

6 $\dfrac{\sqrt{5}}{2}$

7   (a) (i) $\tfrac{14}{9}$ (ii) $\tfrac{13}{162}$ (b) (i) $\tfrac{3}{5}$ (ii) $\tfrac{11}{150}$
       (c) (i) $\tfrac{7}{12}$ (ii) $\tfrac{11}{144}$ (d) (i) $\tfrac{4}{9}$ (ii) $\tfrac{13}{162}$

*****8**   (a) $\tfrac{3}{130}$

$$F(x) = \begin{cases} 0 & x \le 1 \\ \tfrac{1}{26}(x^3 - 1) & 1 \le x \le 3 \\ 1 & x \ge 3 \end{cases}$$

Abbreviated, this is written
$F(x) = \tfrac{1}{26}(x^3 - 1),\ 1 \le x \le 3$

(b)   3; $F(x) = \tfrac{1}{3}(x^2 + x - 2),\ 1 \le x \le 2$

(c) $\tfrac{16}{15}$; $F(x) = \tfrac{16}{15}\left(1 - \dfrac{1}{x^2}\right),\ 1 \le x \le 4$

(d) $\dfrac{3}{\sqrt{2}}$; $F(x) = -\tfrac{1}{2}(2x^2 - 9x + 7),$
      $1 \le x \le 3$

9   (a) $f(x) = 1,\ 1 < x < 2$

(b) $f(x) = 1 - \dfrac{x}{2},\ 0 < x < 2$

(c) $f(x) = \tfrac{2}{15}(x + 1)\ \ 0 < x < 3$

10   (a) $f(x) = \dfrac{x}{4},\ 1 < x < 3$

(b) $\tfrac{13}{6}$ (c) $\tfrac{11}{36}$

11(a) $\dfrac{2}{9}\displaystyle\int_4^m (x)dx = 0.5 \Rightarrow \dfrac{2}{9}\left[\dfrac{x^2}{2}\right]_4^m$

$$= \dfrac{2}{9}\left(\dfrac{m^2}{2} - 8\right) = \dfrac{1}{2}$$

$$\Rightarrow m^2 = 2\left(\dfrac{9}{4} + 8\right),$$
so
$$m = 4.53$$

(b)   4.63   (c) 4.27

12   (a) (i) $f(x) = \tfrac{1}{8},\ 2 \le x \le 10$ (ii) 6
        (iii) $\tfrac{16}{3}$ (iv) $\tfrac{5}{8}$

(b)   (i) $f(x) = \tfrac{1}{3},\ 4 \le x \le 7$ (ii) $\tfrac{11}{2}$
        (iii) $\tfrac{3}{4}$ (iv) $\tfrac{2}{3}$

(c)   (i) $f(x) = \tfrac{1}{8},\ -2 \le x \le 6$ (ii) 2
        (iii) $\tfrac{16}{3}$ (iv) $\tfrac{1}{8}$

(d) (i) $f(x) = \dfrac{1}{5B - 2A}$ $2A \le x \le 5B$

(ii) $\dfrac{2A + 5B}{2}$

(iii) $\frac{1}{12}(5B - 2A)^2$

(iv) $\dfrac{5B - 5}{5B - 2A}$

(e) (i) $f(x) = \dfrac{1}{b^2 - a^2}$ (ii) $\frac{1}{2}(a^2 + b^2)$

(iii) $\dfrac{(b^2 - a^2)^2}{12}$ (iv) $\dfrac{b^2 - 5}{b^2 - a^2}$

13 (a) 7   (b) $F(x) = \dfrac{x - 1}{6}$, $1 \le x \le 7$

14 (a) (i) 6   (ii) 8.5   (iii) $\frac{25}{12}$

(iv) $F(x) = \dfrac{x - 6}{5}$, $6 \le x \le 11$

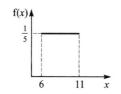

(b) (i) 9   (ii) 9.5   (iii) $\frac{3}{4}$

(iv) $F(x) = \dfrac{x - 8}{3}$, $8 \le x \le 11$

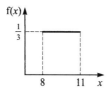

(c) (i) $\frac{10}{3}$ (ii) $\frac{28}{3}$ (iii) $\frac{16}{3}$

(iv) $F(x) = \dfrac{3x - 16}{24}$, $\dfrac{16}{3} \le x \le \dfrac{40}{3}$

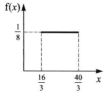

(d) (i) 3   (ii) 11   (iii) $\frac{4}{3}$

(iv) $F(x) = \dfrac{x - 9}{4}$, $9 \le x \le 13$

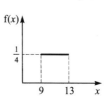

15 (a) $f(x) = \frac{1}{5}$, $2 \le x \le 7$

(b) 4.5, 2.08   (c) $\frac{2}{5}$ (d) $\frac{1}{5}$

* 16 (a) 5, $\frac{4}{3}$ (i) $\frac{1}{2}$ (ii) $\frac{3}{4}$

(b) 11, $\frac{1}{3}$ (i) 1   (ii) $\frac{1}{2}$

(c) 1, $\frac{4}{3}$ (i) $\frac{1}{2}$ (ii) $\frac{1}{10}$

*(d) $E(X) = \frac{1}{2}\{-10 + (-6)\} = -8$,
Var$(X) = \frac{1}{12}\{-6 - (-10)\}^2 = \frac{16}{12} = \frac{4}{3}$

(i) $p(-9 < X < -7) = \frac{2}{4} = \frac{1}{2}$

(ii) $p(-8.5 < X < -7.5) = \frac{1}{4}$

(e) $\dfrac{3A + 5B}{2}, \dfrac{(5B - 3A)^2}{12}$

(i) $\dfrac{5B - 1}{5B - 3A}$ (ii) $\dfrac{2}{5B - 3A}$

(f) $\dfrac{P + Q}{2PQ}, \dfrac{1}{12}\dfrac{(Q - P)^2}{P^2 Q^2}$

(i) $\dfrac{Q(PK^2 - 1)}{P - Q}$ (ii) $\dfrac{P(1 - 10P)}{P - Q}$

17 (a) (i) $\frac{1}{2}$ (ii) 0.788

(b) $F(x) = \dfrac{1}{1.7}$, $0.3 \le x \le 2$

(c) $m = 1.15$

18 (a) $\frac{1}{4}$ (b) 3, $\frac{4}{3}$

(c) (i) 0.961   (ii) $0.417 = \frac{5}{12}$

19 (a) $A = 3$ (b) $A = -2$, $B = 4$

(c) $C = 8$

* 20 (a) $B - A = 5$, $\frac{1}{2}(A + B) = 1 \Rightarrow 2B = 7$,
so $A = -\frac{3}{2}$, $B = \frac{7}{2}$

(b) $p(X < 0) = \dfrac{\frac{3}{2}}{5} = \dfrac{3}{10} = 0.3$

21 (a) 2   (b) $\frac{4}{3}$

Intermediate

2 (a) $\frac{2}{15}$ (i) $\frac{9}{15}$ (ii) $\frac{7}{30}$ (b) $\frac{6}{47}$, $\frac{33}{47}$
(c) $\frac{1}{53}$ (i) $\frac{7}{53}$ (ii) $\frac{46}{53}$

3 (a)(i) $k \displaystyle\int_1^2 (1+x)\mathrm{d}x$

$$= k\left[x + \frac{x^2}{2}\right]_1^2$$

$$= k\{(2+2) - (1+\tfrac{1}{2})\}$$

$$= \frac{5k}{2} = 1, \text{ so } k = \frac{2}{5}$$

*(ii) $\mathrm{E}(X) = \dfrac{2}{5}\displaystyle\int_1^2 (x + x^2)\mathrm{d}x$

$$= \frac{2}{5}\left[\frac{x^2}{2} + \frac{x^3}{3}\right]_1^2$$

$$= \frac{2}{5}\{(2+\tfrac{8}{3}) - (\tfrac{1}{2}+\tfrac{1}{3})\}$$

$$= \frac{23}{15}$$

(iii) $\frac{37}{450}$
(iv) $\frac{1}{8}(x^2 + 2x - 3)$, $1 \le x \le 2$

(b) (i) $\frac{3}{8}$ (ii) $\frac{45}{32}$ (iii) 0.0725
(iv) $\frac{1}{8}(-x^3 + 15x - 14)$, $1 \le x \le 2$

(c) (i) $\frac{3}{2}$ (ii) $\dfrac{3\ln 3}{2} = 1.65$

(ii) 0.284
(iv) $\frac{3}{2}(1 - 1/x)$, $1 \le x \le 3$

4 (a) $F(x) = x^2$, $0 \le x \le 1$
(b) $F(x) = \frac{1}{6}(x + x^2)$, $0 \le x \le 2$
(c) $F(x) = \frac{1}{16}(12x - x^3)$, $0 \le x \le 2$
(d) $F(x) = \frac{1}{7}(x^3 - 1)$, $1 \le x \le 2$

5 (a) (i) 1.54 (ii) 0.495
(b) $F(x) = \frac{1}{6}(x^2 + 3x - 4)$,
$1 \le x \le 2$; 0.514

6 $F(x) = \frac{3}{37}(x^2 + 2x - 3)$, $1 \le x \le 2$;
$\frac{15}{37} + \frac{2}{37}(2x^2 + x - 10)$, $2 \le x \le 3$

7 (a) 3.18   (b) 0.952   (c) 3.53

8 (a) $\frac{8}{21}$ (b) 0.0517

*9 (a) $\displaystyle\int_0^4 (A + Bx)\mathrm{d}x = 1$

$$= \left[Ax + \frac{Bx^2}{2}\right]_0^4 = 4A + 8B = 1$$

$$\int_0^3 (A + Bx)\mathrm{d}x = \frac{1}{2} = \left[Ax + \frac{Bx^2}{2}\right]_0^3$$

$$= 6A + 9B = 1$$

Hence $A = -\frac{1}{12}$, $B = \frac{1}{6}$

10 $A = -\frac{1}{160}$, $B = -\frac{21}{320}$

11 (a) $P = \frac{3}{4}$, $Q = \frac{1}{4}$
(b) $\mathrm{E}(X) = 1$, $\mathrm{Var}(X) = \frac{1}{5}$

12 (b) 2.01 m

13 (a) -2 (b) $\frac{5}{3}$, $\frac{1}{18}$ (c) $\frac{7}{16}$

14 (a) (i) $\frac{3}{46}$ (ii) $\frac{21}{23}$ (iii) 0.314

15 (a) (i) $a = \frac{1}{4}$, $b = \frac{3}{4}$ (ii) $\mathrm{Var}(x) = \frac{1}{48}$
(b) $F(x) = 2$, $\frac{1}{4} \le x \le \frac{3}{4}$

16 (a) $k = \frac{1}{4}$
(b) (i) 1 (ii) 3 (iii) $\frac{4}{3}$ (iv) $\frac{5}{6}$

*17 (a) $\frac{1}{2}(A + A + 5) = 3$,
$\therefore 2A = 6 - 5 = 1$,
$\therefore A = \frac{1}{2}$
(b) $\mathrm{p}(X < \frac{4}{3}) = \dfrac{\frac{4}{3} - 1}{5} = \frac{1}{5}\left(\dfrac{8-3}{6}\right) = \frac{1}{6}$.

(c) $\mathrm{E}(X) - \mathrm{Var}(X) = 3 - \frac{1}{12}(5\frac{1}{2} - \frac{1}{2})^2$
$= 3 - \frac{25}{12} = \frac{11}{12}$
$\mathrm{p}(X > \{\mathrm{E}(X) - \mathrm{Var}(X)\})$
$= \mathrm{p}(X > \frac{11}{12}) = \dfrac{5\frac{1}{2} - \frac{11}{12}}{5} = \frac{11}{12}$

18 (a) $\dfrac{1}{2.4}$ (b) $\mathrm{f}(x) = \dfrac{1}{2.4}$, $0.6 \le x \le 3$

(c) $F(x) = \dfrac{x - 0.6}{2.4}$, $0.6 \le x \le 3$

(d) 0.583

19 $f(c) = \dfrac{1}{2\pi}$, $2\pi < c < 4\pi$; $E(C) = 3\pi$

20 (a)

(b) (i) 5, $\frac{16}{3}$ (ii) $\frac{1}{4}$ (iii) $\frac{1}{24}$ (iv) $\frac{11}{24}$ (v) 2

21 14, 12

Advanced

1 (a) $\frac{1}{800}$ (b) 75.1 (c) 8.95 (d) 0.516

2 (a) $\frac{1}{400}$ (b) 20 (c) $\frac{3}{8}$ (d) $\frac{3}{4}$

3 (a) $a = \frac{1}{5}, b = \frac{1}{10}$ (b) $\frac{17}{15}$
  (c) $\frac{71}{225} = 0.316$

*4 (a) (i) $\frac{1}{18}$

  (ii) $F(b) = \displaystyle\int_1^b \frac{1}{18}(3x^2 - 2x)\mathrm{d}x$

  $= \frac{1}{18}[x^3 - x^2]_1^b$

  $= \frac{1}{18}(b^3 - b^2)$

  Hence

  $$F(x) = \begin{cases} 0 & x < 1 \\ \frac{1}{18}(x^3 - x^2) & 1 \le x \le 3 \\ 1 & x \ge 3 \end{cases}$$

  (iii) 0.226 (iv) $\frac{7}{9}$

5 (a) $-\frac{1}{44}$ (b) $\frac{7}{3}$ (c) $\frac{3}{2}$

6 (a) (i) $\frac{6}{49}$ (ii) $\frac{15}{14}$ (iii) 0.368
  (iv) $F(x) = \frac{6}{49}(4x - x^3/3)$,
    $0 \le x \le 1$, and
    $\frac{22}{49} + \frac{9}{49}(x^2 - 1)$, $1 \le x \le 2$
  (c) 0.0122

7 (a) $\dfrac{2}{\pi - 4} = -2.33$

(b) $\dfrac{\pi + 4}{4} = 1.79$

(c) $F(x) = \dfrac{2}{\pi - 4}(x - 2 + 2\cos x)$,
  $0 \le x \le \dfrac{\pi}{2}$

8 (i) $\frac{1}{2}$
  (ii) $F = 0$, for $x < 0$, $F = x - \frac{1}{4}x^2$, for
    $0 \le x \le 2$, $F = 1$ for $x > 2$;
    $f(y) = 2y$

9 (a) 0, 40
  (b) (i) 20 (ii) $\frac{1600}{3}$ (iii) $\frac{200}{3}$ (iv) 0.25

10 17, $33\frac{1}{3}$

11 (a) 13
  (b) 27
  (c) 196
  (d) $f(y) = \frac{1}{18}$, $4 < y < 22$
  (e) $\frac{1}{9}$

12 (a) 2, 12 (b) $\frac{1}{10}$ (c) 31, 208

*13 (a) $\frac{1}{2}(A + B) = 3$, so $A + B = 6$
    $B - A = \dfrac{1}{1/B} = B$, hence $A = 0$,
    $B = 6$
  (b) $\mathrm{Var}(X) = \frac{1}{12} \times 36 = 3$
  (c) If $Y = 3X$, then $E(Y) = 3 \times 3 = 9$
  (d) $\mathrm{Var}(Y) = 9\,\mathrm{Var}(X) = 9 \times 3 = 27$

14 (a) $\frac{1}{4}\mathrm{e}^{-t/4}$
  (b) (i) 4 minutes
    (ii) $F(t) = 1 - \mathrm{e}^{-t/4}$
    (iii) 2.77 minutes
  (c) 0.287

Revision

1 (a)

$A = \frac{8}{3}$; no

(b)

$A = 1$; yes

(c)

$A = 1$; yes

(d)

$A = 3$; no

*4 *(a) $E(X) = \dfrac{1}{3}\displaystyle\int_0^1 (4x^2 + x)dx$

$$= \frac{1}{3}\left[\frac{4x^3}{3} + \frac{x^2}{2}\right]_0^1$$

$$= \frac{1}{3}\left(\frac{4}{3} + \frac{1}{2}\right)$$

$$= \frac{11}{18}$$

$$E(X^2) = \frac{1}{3}\int_0^1 (4x^3 + x^2)dx$$

$$= \frac{1}{3}\left[x^4 + \frac{x^3}{3}\right]_0^1$$

$$= \frac{1}{3}\left(1 + \frac{1}{3}\right)$$

$$= \frac{4}{9}$$

$$\text{Var}(X) = \frac{4}{9} - \left(\frac{11}{18}\right)^2 = \frac{23}{324}$$

(b) $\frac{8}{3}, \frac{13}{18}$

5 (a) (i) $\frac{1}{6}$ (ii) 1.30 (iii) $\frac{23}{81}$

(b) (i) $\frac{4}{3}$ (ii) 0.581 (iii) $\frac{13}{162}$

6 (a) $F(x) = \frac{3}{2}(1 - 1/x)$, $1 \le x \le 3$

(b) $F(x) = \frac{1}{29}(36x - x^3 - 35)$,
$1 \le x \le 2$

(c) $F(x) = \frac{1}{27}(12x - x^3 + 11)$,
$-1 \le x \le 2$

7 (a) $\frac{3}{2}$ (b) 2.62   (c) 2.52
(d) 0.375   (e) 0.16   (f) 2.37

8 (a) $\dfrac{1}{5.625} = 0.178$

(b) 0.711

(c) 1 hour 29 minutes

9 (a) 0.448   (b) 0.168
(c) $f(x) = \frac{1}{5}(10 - 2x)$, $2 < x < 3$
(d) 2.47

10 (a) 222 kg
(b) $F(x) = \frac{2}{3}(4x - 5x^2/2)$, $0 \le x \le 1$

11 (a) $\frac{1}{12}$ (b) 1.61   (c) 0.161

12 (a) $f(x) = \frac{1}{5}$, $15 < x < 20$; 17.5; $\frac{25}{12}$
(i) $\frac{3}{5}$ (ii) 0.3
(b) $f(x) = \frac{1}{6}$, $-4 < x < 2$; -1; 3
(i) $\frac{1}{3}$ (ii) $\frac{5}{6}$ (iii) $\frac{5}{6}$
(c) $f(x) = \frac{1}{6}$, $B - 1 < x < B + 5$;
$B + 2$; 3
(i) $\frac{1}{2}$ (ii) $\frac{5}{6}$ (iii) $\frac{1}{2}$
(d) $f(x) = \frac{1}{20}$, $A - 8 < x < A + 2$;
$A + 2$; $\frac{100}{3}$
(i) $\dfrac{A + 8}{20}$ (ii) $\dfrac{A + 6}{20}$

13 (a) $\dfrac{1}{b - a}$
(b) $a = 3, b = 6, k = \frac{1}{3}$
(c) 5.5

14 (a) $A = \frac{1}{2}B$ (b) $A = 4, B = 8$

15 $P = 1, Q = 6$.

16 $E(Y) = 18$, $\text{Var}(Y) = \text{Var}(X) = 12$

17 (a) f(x)

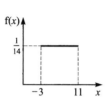

(b) 2, $\frac{25}{3}$

(c) (i) 0.3   (ii) $\frac{1}{2}$ (iii) $\frac{1}{10}$ (iv) 0.15
(d) $A = 4.5$

18 $A = -3, B = 11, C = 7.5, D = -1,$
     $E = 6$

f(x)

4.2

Basic

1 (a) 0.8413   (b) 0.7257   (c) 0.9772
   (d) 0.5897   (e) 0.8264   (f) 0.9147
   (g) 0.7177   (h) 0.8480   (i) 0.9905
   (j) 0.9971

2 (a) 0.2420   (b) 0.1841   (c) 0.1151
   (d) 0.0202   (e) 0.1914   (f) 0.3874
   (g) 0.0088   (h) 0.0231   (i) 0.0067
   (j) 0.4741

3 (a) 0.6002   (b) 0.8520   (c) 0.7767
   (d) 0.6901   (e) 0.9931   (f) 0.9340
   (g) 0.9814

4 (a) 0.3821   (b) 0.2119   (c) 0.0808
   (d) 0.0495   (e) 0.0832   (f) 0.0029
   (g) 0.1471

5 (a) 0.4134   (b) 0.6524   (c) 0.5812
   (d) 0.5872   (e) 0.7866   (f) 0.5985
   (g) 0.8334

6 (a) 0.1151   (b) 0.2225   (c) 0.0855
   (d) 0.2013   (e) 0.2116   (f) 0.1002
   (g) 0.0896

*7 *(a) p($-0.8 < X < -0.4$)
              $= \Phi(0.8) - \Phi(0.4)$
              $= 0.7881 - 0.6554 = 0.1327$
   (b) 0.1464   (c) 0.1647   (d) 0.1422
   (e) 0.1603   (f) 0.1026   (g) 0.1428

*8 (a) 0.3830   (b) 0.1586   (c) 0.7154
   (d) 0.8858   (e) 0.9886
   *(f) p($|Z| > 0.35$) $= 2\{1 - \Phi(0.35)\}$
              $= 2(1 - 0.6368)$
              $= 0.7264$
   (g) 0.4122   (h) 0.2584   (i) 0.1010
   (j) 0.0444

9 (a) 0.6772   (b) 0.1949   (c) 0.0537
   (d) 0.5540   (e) 0.8330   (f) 0.1037
   (g) 0.6348   (h) 0.1441   (i) 0.2542
   (j) 0.7207

10 (a) 0.3085   (b) 0.9861   (c) 0.1151
   (d) 0.7881   (e) 0.5403

11 (a) 0.1464   (b) 0.9876   (c) 0.8904
   (d) 0.2302   (e) 0.0668

12 (a) 56.3   (b) 53.3   (c) 60.3
   (d) 51.46   (e) 56.96

Intermediate

1 (a) 4   (b) 9   (c) 5.29   (d) 0.0225

2 (a) 1.1025   (b) 6.25
   (c) 1.225×10^{-5} (d) 64

3 (a) 5.6   (b) 150   (c) 75   (d) 0.25

4 (a) 43.12   (b) 287   (c) 12   (d) −2.4

5 6

6 (a) 0.0303   (b) 0.468

7 56.6, 12.9

8 25.8%

*9 $p(14.2 < D < 15.8) = 1 - 0.038 = 0.962$

$$\Phi\left(\frac{0.8}{\sigma}\right) - \left\{1 - \Phi\left(\frac{0.8}{\sigma}\right)\right\} = 0.962,$$

$$2\Phi\left(\frac{0.8}{\sigma}\right) = 1.962,$$

$$\Phi\left(\frac{0.8}{\sigma}\right) = 0.981 = \Phi(2.074)$$

so $\sigma = \dfrac{0.8}{2.074}$, variance $= \sigma^2 = 0.149$

10 30.96 cm–31.04 cm

11 (a) 0.7734   (b) 0.8413   (c) 0.0668
   (d) 0.5328   (e) 0.1865

12 (a) (i) 0.0062   (ii) 0.3858   (iii) 0.3085
   (b) 0.0203

13 (a) $\mu = 20$ (b) 0.8904

14 $\mu = 21.3$

15 $\mu = 12, \sigma^2 = 4$

Advanced

*1 Proportion $> 7.1 = \frac{15}{120} = 0.125$
   Proportion $< 6.7 = \frac{10}{120} = 0.0833$

$$1 - \Phi\left(\frac{7.1 - \mu}{\sigma}\right) = 0.125,$$

$$\Phi\left(\frac{7.1 - \mu}{\sigma}\right) = 0.875 = \Phi(1.15),$$

so $7.1 - \mu = 1.15\sigma$

$$1 - \Phi\left(\frac{\mu - 6.7}{\sigma}\right) = 0.0833,$$

$$\Phi\left(\frac{\mu - 6.7}{\sigma}\right) = 0.9167 = \Phi(1.384),$$

so $\mu - 6.7 = 1.384\sigma$
Adding, $0.4 = 2.534\sigma,$

so $\sigma^2 = \left(\dfrac{0.4}{2.534}\right)^2 = 0.0249$

$\mu = 6.7 + 1.384(0.1578) = 6.92$ m
Proportion longer than

$$7.2 \text{ m} = 1 - \Phi\left(\frac{7.2 - 6.92}{0.1578}\right)$$

$$= 1 - \Phi(1.774) = 0.038$$

So expected number of jumps longer
than $7.2\,\text{m} = (20 \times 0.038) = 4.56 \approx 5$
to the nearest whole number

2 246.3 g

3 (a) 0.1425   (b) 497 ml

4 12.1 g

5 2.938 kg–3.065 kg

6 338.1 g–341.9 g; 13.8

7 (a) 0.0228   (b) 18   (c) £57.75

8 (a) 3.16×10^{-4}
   (b) (i) 0.00157   (ii) 0.0288

9 (i) 10.825, 3.258   (ii) 303/304/305
   (iii) 18.4 units

Revision

1 (a) 95.96   (b) 101.8   (c) 108.4
   (d) 107.1   (e) 98.48

2 (a) 33.56   (b) 58.30   (c) 48.99
   (d) 52.02   (e) 60.26

3 (a) (i) 0.0401   (ii) 0.000 135
   (b) 8.62×10^{-4}

4 (a) $\mu = 47, \sigma = 11.9$
      (i) 0.5995   (ii) 0.0551
      (iii) 0.3212   (iv) 0.3256
   (b) 2.40

*5 $1 - \Phi\left(\dfrac{1.1}{\sigma}\right) = 0.22$, gives $\sigma = 1.42$

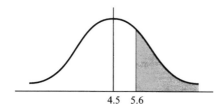

(a) $p(Y < 3.3) = 1 - \Phi\left(\dfrac{1.2}{1.42}\right)$

$= 1 - 0.8009 = 0.1991$

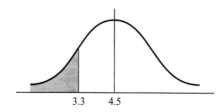

(b) $p(3.8 < Y < 5.1)$

$= \Phi\left(\dfrac{0.6}{1.42}\right) - \left\{1 - \Phi\left(\dfrac{0.7}{1.42}\right)\right\}$

$= 0.6639 + 0.689 - 1 = 0.3529$

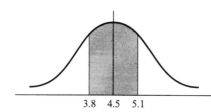

(c) $p(|Y - 4.5| > 0.8)$

$= 1 - \Phi\left(\dfrac{0.8}{1.42}\right) + 1 - \Phi\left(\dfrac{0.8}{1.42}\right)$

$= 2 - 1.4266 = 0.5734$

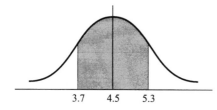

6 (a) 13.43   (b) 0.0278   (c) 0.585
(d) 0.2321

7 (a) $\mu = 10.1, \sigma^2 = 4.85$ (b) 0.9634

8 (a) 0.0258   (b) 0.1644   (c) 3

9 (a) $\mu = 0.997\,01$, $\sigma^2 = 6.78 \times 10^{-6}$
(b) 2.77%

10 $\sigma = 0.0291$; 8.45%

11 (a) 103   (b) 104.7   (c) 94
(d) 98   (e) 93

12 (a) -3 (b) -0.5 (c) -1.8
(d) 2.32   (e) 0.7

4.3

Basic

1 (a) (i) 0.9633   (ii) 0.8144
(b) (i) 0.2281   (ii) 0.0680
(c) (i) 0.9855   (ii) 0.8478
(d) (i) 0.0152   (ii) 0.2819
(e) (i) 0.8864   (ii) 0.9154

2 (a) 0.1791   (b) 0.6950   (c) 0.2377
(d) 0.6950   (e) 0.4433   (f) 0.3638

3 (a) 0.1823   (b) 0.2290   (c) 0.3779
(d) 0.7181

*4 Continuity corrections used

(a) $p(X > 312.5) = 1 - \Phi\left(\dfrac{12.5}{\sqrt{250}}\right)$

$= 1 - \Phi(0.791)$

$= 0.2145$

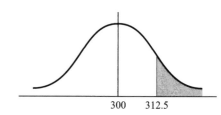

(b) p(297.5 < X < 303.5)

$$= \Phi\left(\frac{3.5}{\sqrt{250}}\right) - \left\{1 - \Phi\left(\frac{2.5}{\sqrt{250}}\right)\right\}$$

$$= 0.5875 + 0.5628 - 1 = 0.1503$$

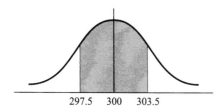

297.5   300   303.5

(c) $p(X < 289.5) = 1 - \Phi\left(\frac{10.5}{\sqrt{250}}\right)$

$$= 1 - 0.7467$$

$$= 0.2533$$

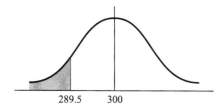

289.5   300

(d) $p(X=303)=\Phi\left(\frac{3.5}{\sqrt{250}}\right)-\Phi\left(\frac{2.5}{\sqrt{250}}\right)$

$$= 0.5875 - 0.5628 = 0.0247$$

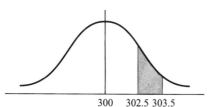

300   302.5 303.5

5   (a) 0.4286   (b) 0.1343   (c) 0.2771

6   (a) 0.0649   (b) 0.7821   (c) 0.4552
(d) 0.0968; 0.0057

7   (a) 0.0583   (b) 0.1343

8   (a) (i) 0.9145   (ii) 0.9723
(iii) 0.1578   (iv) 0.0604
(v) 0.6758

(b) (i) 0.9825   (ii) 0.7789
(iii) 0.0123   (iv) 0.8824
(v) 0.3074

9   (a) (i) 0.0287   (ii) 0.1841   (iii) 0.6612
(b) 2

10 (a) 0.0140   (b) 0.0055; 0.8477

11 (a) (i) 0.0485   (ii) 0.0461
(b) (i) 0.0519   (ii) 0.0484

12 (a) 0.0285   (b) 0.0297

Intermediate

1   407

2   (a) 0.0586   (b) 0.2296   (c) 0.0107

3   (a) 100
(b) (i) 0.0495   (ii) 0.0354   (iii) 0.0461

*** 4**   (a) Poisson $\lambda = 21 \times 2 = 42$,

$$p(X = 48) = \frac{e^{-42}42^{48}}{48!} = 0.0382$$

Normal (42, 42),

$$p(X = 48) = \Phi\left(\frac{6.5}{\sqrt{42}}\right) - \Phi\left(\frac{5.5}{\sqrt{42}}\right)$$

$$= 0.0400$$

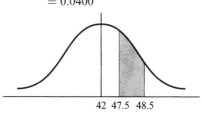

42 47.5 48.5

(b) (i) $p(X < 35) = 1 - \Phi\left(\frac{7.5}{\sqrt{42}}\right)$

$$= 1 - 0.8763 = 0.1237$$

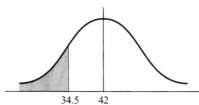

34.5   42

(ii) $p(X \geq 52) = 1 - \Phi\left(\dfrac{9.5}{\sqrt{42}}\right)$

$= 0.0713$

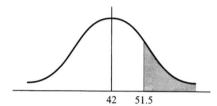

5 (a) 0.0326   (b) 0.0713

6 (a) 0.1359, 0.1384
   (b) 0.4834, 0.4875

7 (a) (i) 0.0642   (ii) 0.0668
   (b) (i) 0.0681   (ii) 0.0634

8 (a) (i) 0.0620   (ii) 0.5627   (iii) 0.8344
   (b) (i) 0.2348   (ii) 0.1570

9 (a) (i) 0.0599
         (ii) 0.0628; 4.84% higher
   (b) (i) 0.0426
         (ii) 0.0402; 5.63% lower

10 (a) (i) 0.0569   (ii) 0.0295   (iii) 0.0137
    (b) 0.0575   0.0307   0.0147;
       0.0618   0.0282   0.0102

11 (a) 0.00199   (b) 0.1612

12 (a) 0.9334   (b) 0.4830

13 (a) Poisson 0.00962, normal 0.0099
    (b) 0.0356

14 (a) Poisson 0.0614, normal 0.0614
    (b) Poisson 0.0903, normal 0.0858

15 (a) 0.0559   (b) 6.25 × 10⁻⁴

Advanced

1 (a) (i) 0.0432
         (ii) binomial 0.0415, normal
              0.0401
         (iii) 0.9105

(b) (i) binomial 0.0897, normal
       0.0866
   (ii) 0.0890

*2 (a) Binomial $= (0.98 + 0.02)^{100}$
      $p(X \geq 3)$
      $= 1 - \{p(X = 0) + p(X = 1)$
      $\quad + p(X = 2)\}$
      $= 1 - \{0.98^{100} + 100(0.98)^{99}$
      $\quad \times (0.02) + {}^{100}C_2(0.98)^{98}(0.02)^2\}$
      $= 0.3230$
   (b) Normal $= N(2, 1.96)$

   $$p(X > 2.5) = 1 - \Phi\left(\frac{2.5 - 2}{\sqrt{1.96}}\right)$$

   $$= 0.3606$$

   (c) Poisson $= Po(2)$
      $p(X \geq 3) = 1 - \{p(X = 0)$
      $\quad\quad\quad + p(X = 1) + p(X = 2)\}$

      $$= 1 - e^{-2}\left(1 + 2 + \frac{2^2}{2!}\right)$$

      $$= 1 - 5e^{-2}$$

      $$= 0.3233$$

      Normal discrepancy $= 11.6\%$, p is
      too small, but Poisson is very
      good because p is small.

3 (a) (i) 0.0508   (ii) 0.7817
   (b) 0.2352

4 (a) Binomial 0.0100, normal 0.0102
   (b) Binomial 0.0215, normal 0.0215
   (c) Binomial 0.0229, normal 0.0224
   (d) Binomial 0.0215, normal 0.0378
      p(25 sixes) = 0.0014 (normal),
      0.000 627 (binomial)
      p(20 'ones') = 0.0061 (normal),
      0.007 44 (binomial), so 20 'ones'
      are more likely

5 (a) (i) 0.1044   (ii) 0.7489   (iii) 0.8093
   (b) 0.0418   (c) 0.8506

6 (a) 0.0293
   (b) (i) 0.1922   (ii) 0.2312

7 (a) (i) 0.0378 (ii) 0.0103
   (b) (i) 0.0485 (ii) 0.4830

8 (i) 0.747
   (ii) Shortsightedness is often inherited.
   (iii) 0.0032

9 (i) 0.0278 (ii) 0.0279

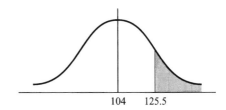

Revision

1 (a) (i) 0.00364 (ii) 0.00360
   (b) (i) 0.00671 (ii) 0.0123
   (c) (i) 0.0327 (ii) 0.0347
   (d) (i) 0.0654 (ii) 0.0693

2 (a) (i) 0.6992 (ii) 0.4769 (iii) 0.0087
   (b) (i) 0.0259 (ii) 0.8920
      (iii) 0.8884 (iv) 0.3766

3 (a) (i) 0.1404 (ii) 0.8786
      (iii) 1.215×10^{-4}
   (b) (i) 9.13×10^{-6} (ii) 0.00975
      (iii) 0.9858
   (c) (i) 0.0476 (ii) 0.1911

4 (a) (i) 0.0624 (ii) 0.0263
   (b) (i) 0.0185 (ii) 0.9334

5 (a) 0.0086 (b) 0.0461 (c) 0.9504

*6 Po(8)

   (a) (i) $p(X = 3) = \dfrac{e^{-8} \cdot 8^3}{3!} = 0.0286$

   (ii) $p(X \leq 4) = p(X = 0) + p(X = 1)$

$$+ p(X = 2) + p(X = 3)$$
$$+ p(X = 4)$$

$$= e^{-8}\left(1 + 8 + \frac{8^2}{2!} + \frac{8^3}{3!} + \frac{8^4}{4!}\right)$$

$$= 0.0996$$

   (b) For 0600–1900,
      $\lambda = 104 \Rightarrow X = N(104, 104)$

$$p(X > 125) = 1 - \Phi\left(\frac{21.5}{\sqrt{104}}\right)$$
$$= 0.0176$$

7 (a) (i) 0.0669 (ii) 0.00992
   (b) (i) 0.9395 (ii) 0.0032

8 (a) 0.0258 (b) 0.0674

9 (a) Poisson 0.007 57, normal 0.0112:
      large error, $V = 3$ in tail of
      distribution, λ small
   (b) Poisson 0.067 09, normal 0.0774:
      large error, $V < 6$ near tail and λ
      small
      Normal calculation is much
      quicker.
   (c) Poisson 0.061 87, normal 0.0637:
      small error, $W = 37$ in centre of
      distribution, λ large
   (d) Poisson 0.002 325, normal 0.0044:
      large error. Normal is very much
      quicker to calculate.

10 (a) Binomial 0.2204, normal 0.2194:
      small error, value from centre of
      distribution
   (b) Binomial 0.5730, normal 0.5753:
      small error, normal much quicker
   (c) Binomial 0.0153, normal 0.0158:
      small error, both easy to calculate
   (d) Binomial 0.0168, normal 0.0227:
      large error, binomial a very long
      calculation

11 (a) 3.80×10^{-4}
   (b) 0.2874
   (c) 0.1762

12 (a) 0.1759
   (b) Binomial 0.214, normal 0.229
   (c) 0.0588

4.4

Basic

1 (a) (i) 0.6613 (ii) 0.9173 (iii) 0.2028
(iv) 0.1339 (v) 0.7105 (vi) 0.8343
(vii) 0.1339 (viii) 0.2441
(b) (i) 0.7105 (ii) 0.3490 (iii) 0.3089
(iv) 0.7473 (v) 0.4195
(vi) 0.1125 (vii) 0.5290
(viii) 0.6178
(c) (i) 0.9084 (ii) 0.7105 (iii) 0.1525
(iv) 0.4013 (v) 0.7645 (vi) 0.9430
(vii) 0.3327 (viii) 0.5630

2 (a) N(65, 1.52) (b) N(30, 1.4)
(c) N(30, 3.24) (d) N(40, 4)
(e) N(75, 9.81) (f) N(5, 0.34)
(g) N(−2.5, 0.085)

3 (a) N(52, 40) (b) N(120, 113)
(c) N(9, 18) (d) N(−16, 45)
(e) N($5, \frac{17}{4}$) (f) N(47, 15)
(g) N(24, 129) (h) N(109, 70)
(i) N(10, 117) (j) N(21, 3.47)

*4 (a) $X = $ N(800, 64),
$p(X > 815) = 1 − \Phi(\frac{15}{8}) = 0.0303$

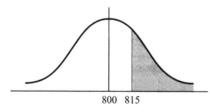

800 815

(b) $X = $ N(6400, 512),
$$p(X < 6360) = 1 − \Phi\left(\frac{40}{\sqrt{512}}\right)$$
$$= 0.0385$$

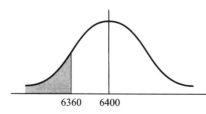

6360   6400

5 0.0787; 0.5202

6 (a) (i) N(55, 41) (ii) N(−5, 41)
(iii) N(5, 41) (iv) N(80, 116)
(v) N(10, 544) (vi) N(65, 104)
(b) (i) 0.3196 (ii) 0.3608
(iii) 0.6464 (iv) 0.0707

7 (a) 2, 30 (b) 0.0721

8 $\mu = 10$, $\sigma^2 = 4$; $X = $ N(10, 4),
$Y = $ N(40, 16)

9 $k = 5$

10 (a) 0.6814 (b) 0.0408

11 (a) (i) 0.0082 (ii) 0.0228
(b) $(2L + 2S) = $ N($44, \frac{5}{8}$)

12 (a) 0.0910 (b) 0.0055

Intermediate

1 (a) 0.7340 (b) 0.8716 (c) 0.0688
(d) 0.8942 (e) 0.9856 (f) 0.4612

*2 (a) 0.027 (b) 0.6675 (c) 0.027
*(d) $2Q − P = $ N(0, 22.6),
so $p(2Q − P > 0) = 0.5$

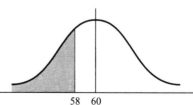

$\sigma = \sqrt{22.6}$

0

*(e) $P + Q = $ N(60, 6.73),
$p(P + Q < 58)$
$$= 1 − \Phi\left(\frac{2}{\sqrt{6.73}}\right) = 0.2203$$

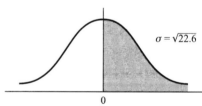

58 60

3 (a) 0.6469   (b) 0.6436   (c) 0.5375

4 (a) 0.2119   (b) 0.068   (c) 0.8882

5 (a) (i) 0.9772   (ii) 0.1319
 (b) 0.0127   (c) 1 game

6 (a) $\mu = 50$, $\sigma = 7$ (b) 0.9394

7 (a) 0.8664   (ii) 0.9813   (b) 0.7364

8 (a) 0.0787   (b) 0.6818

9 (a) p = N(1014, 0.14)
 (b) (i) 0.6657   (ii) 0.2405

10 (a) (i) N(9, 0.432)   (ii) N(−1, 0.15)
 (iii) N(−1, 0.15)
 (iv) N(0.2, 0.000 161 6)
 (b) (i) 0.798 6   (ii) 0.0049
 (iii) 0.9953

11 (a) 0.9902   (b) $L = N(990, 135)$
 (c) $M_1 = 980$ g, $M_2 = 1000$ g

*12 (a) (i) p(R > 4.2)

$$= 1 - \Phi\left(\frac{1.2}{1}\right)$$

$$= 1 - 0.8849 = 0.1151$$

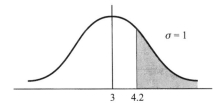

$\sigma = 1$

3   4.2

(ii) $p(R > 0.25) = 1 - \Phi\left(\dfrac{0.15}{\sqrt{\dfrac{1}{30}}}\right)$

$$= 1 - 0.7944$$

$$= 0.2056$$

{for 1 day $R = N(\frac{1}{10}, \frac{1}{30})$}

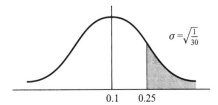

$\sigma = \sqrt{\dfrac{1}{30}}$

0.1   0.25

(b) $1 - \Phi\left(\dfrac{0.2}{\sqrt{\dfrac{1}{30}}}\right) = 1 - \Phi(1.095)$

$$= 0.1367$$

Hence no. of days $= 30 \times 0.1367 \approx 4$

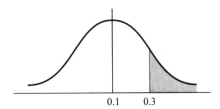

0.1   0.3

13 (a) (i) 0.1562   (ii) 0.1241
 (b) 15 races

14 (a) 0.0228   (b) 0.0416
15 (a) 0.7257   (b) 0.7734
 (c) 0.0827   (d) 0.8185

Advanced

1 (a) 0.2266   (b) 0.1573   (c) 0.0062

*2 $X + Y = N\left(\dfrac{7\mu}{3}, \dfrac{25\sigma^2}{9}\right)$

$$p(X + Y < 45) = \Phi\left(\dfrac{45 - \dfrac{7\mu}{3}}{\dfrac{5\sigma}{3}}\right)$$

$$= 0.9772 = \Phi(2)$$

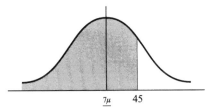

$\dfrac{7\mu}{3}$ 45

so $45 - \dfrac{7\mu}{3} = \dfrac{10\sigma}{3} \Rightarrow 135 - 7\mu = 10\sigma$

$$(1)$$

$$p(X + Y < 40) = \Phi\left(\dfrac{40 - \dfrac{7\mu}{3}}{\dfrac{5\sigma}{3}}\right)$$

$$= 0.8413 = \Phi(1)$$

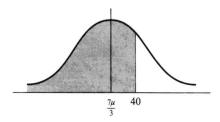

so $40 - \dfrac{7\mu}{3} = \dfrac{5\sigma}{3} \Rightarrow 120 - 7\mu = 5\sigma$ (2)

From (1) and (2), $\mu = 15$, $\sigma = 3$

$Y - X = N(5, 25)$, so

$$p(Y - X < 0) = 1 - \Phi(\tfrac{5}{5})$$

$$= 1 - 0.8413 = 0.1587$$

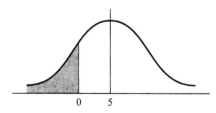

3 (a) (i) 0.0082 (ii) 0.0317   (b) 0.071

4 (a) 0.2025   (b) 0.13   (c) 0.8829

5 (a) (i) 0.1587 (ii) 0.9873
    (b) 0.0228

6 (a) (i) 0.0913 (ii) 0.0038
    (b) 0.085

7 (a) (i) 0.0649 (ii) 0.6919 (iii) 0.9214
      (iv) 0.223
    (b) 82   (c) 0.108

8 (i) 0.1078 (ii) 0.2399

9 (i) 0.1056 (ii) 0.709 (iii) 0.3043
    (iv) 0.0763;
    (a) Performance may be similar on
      different mountains or suffer from
      tiredness as more mountains are
      climbed.
    (b) Team members affect one
      another's performance.

Revision

1 (a) (i) N(450, 720) (ii) N(600, 300)
      (iii) N(240, 1820)
      (iv) N(−165, 260)
      (v) N(300, 1820)
    (b) (i) 0.8243 (ii) 0.4584 (iii) 0.6516
      (iv) 0.9687 (v) 0.2412

2 $\mu = 64.0$

3 $k = 3$

4 (a) (i) 0.0199 (ii) 0.9525
    (b) (i) 9 (ii) 6

5 $\mu = 40.9$, $\sigma = 7.10$; 0.316

6 (a) $\mu = 10, \sigma^2 = 4$
    (b) 0.9568
    (c) 0.7120

*7 (a) If $4B - 3A = N(260, 1390)$, then
      $(16 - 3)\mu = 260 \Rightarrow \mu = 20$ and
      $$\left(\dfrac{16 \times 7}{3} + 9\right)\sigma^2 = 1390 \Rightarrow \sigma^2 = 30$$
      Hence $A = N(20, 30)$,
      $B = N(80, 70)$

    (b) $2A + B = N\left(6\mu, \dfrac{19\sigma^2}{3}\right)$

      $$= N(120, 190)$$

      $$p(2A + B < 130) = \Phi\left(\dfrac{10}{\sqrt{190}}\right)$$
      $$= \Phi(0.725)$$
      $$= 0.7657$$

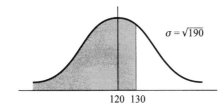

$\sigma = \sqrt{190}$

120 130

10 (a) (i) N(0.28, 0.0012)
       (ii) N(−0.12, 0.0012)
       (iii) N(−0.84, 0.0258)
       (iv) N(0.025, 0.00001078)
   (b) (i) 0.1933
       (ii) 0.1241
       (iii) 0.9829
       (iv) 0.9835

8 (a) 0.0385   (b) 0.8772   (c) 0.8369

11 (a) (i) 0.0107   (ii) 0.9108
   (b) 0.8159

9 (a) 0.1333   (b) 0.0684   (c) 0.7688
   (d) 0.0584

12 (a) (i) 0.7734   (ii) 0.5160
   (b) (i) 0.0898   (ii) 1.05 cm

Chapter 5 Correlation and regression

Basic

1 (a) (7, 10) (b) (4.5, 4) (c) (4.7, 7)
   (d) (15, 5) (e) (13.75, 4)

2 (a) (5.5, 10) (b) (9, 14)
   (c) (12.7, 12.4)

3 (a) (11, 8); $y = 10.4$
   (b) (4.5, 4); $x = 4.98$
   (c) (6.5, 10.4); $y = 13.9$
   (d) (22.5, 41.3); $x = 37.8$

4 (a) (i) 5.83 (ii) 2.92 (iii) 11.7
   (b) (i) -21.3 (ii) 72.9 (iii) 6.25
   (c) (i) -32.9 (ii) 46.7 (iii) 25.1
   (d) (i) 34.3 (ii) 81.3 (iii) 14.8

* 5 (a) $r = 0.9999 \approx 1$, perfect positive
   correlation (nearly!)
   *(b) $\Sigma x = 21$, $\Sigma y = 58$, $\Sigma x^2 = 91$,
   $\Sigma y^2 = 636$, $\Sigma xy = 217$

   $$r = \frac{6 \times 217 - 21 \times 58}{\sqrt{(546 - 441)} \times \sqrt{(3816 - 3364)}}$$

   $$= \frac{84}{\sqrt{105}\sqrt{452}} = 0.386$$

   $r = 0.386$, low positive correlation

   (c) $r = 0.128$, very low positive
   correlation
   (d) $r = -1$, perfect negative
   correlation
   (e) $r = -0.794$, high negative
   correlation

6 (a) (i) $r = 0.992$
   (ii) $y = -12.4 + 4.6x$,
   $x = 2.73 + 0.214y$

(b) (i) $r = -0.594$
   (ii) $y = 18.9 - 1.28x$
   $x = 9.24 - 0.276y$
(c) (i) $r = 0.994$
   (ii) $y = -3.51 + 1.37x$,
   $x = 2.66 + 0.719y$

7 (a) 0.894 (b) -0.669 (c) -0.730
   (d) 0.9 (e) 0.344

8 (a) $y = -3.25 + 2.25x$,
   $x = 1.55 + 0.432y$; $r = 0.985$
   (b) $y = 30.5 - 1.14x$,
   $x = 26.2 - 0.837y$; $r = -0.976$
   (c) $y = 10.6 + 1.34x$,
   $x = -7.71 + 0.741y$; $r = 0.995$
   (d) $y = 12.3 - 0.0962x$,
   $x = 14.4 - 0.126y$; $r = -0.110$

9 $a = 2$, $b = \frac{1}{2}$, $c = 1$, $d = \frac{2}{3}$; $r = 0.577$

* 10 *(a) Equation of y on x is

   $$y - \frac{66}{7} = \frac{7 \times 369 - 42.5 \times 66}{7 \times 264.75 - (42.5)^2}$$
   $$\times \left(x - \frac{42.5}{7} \right)$$

   Hence

   $$y - 9.4286 = \frac{-222}{47}(x - 6.071),$$

   so $y = 38.1 - 4.72x$,
   $x = 7.42 - 0.143y$
   (b) (i) $y = 2.7$ (ii) $x = 6.82$
   (c) $r = -0.822$

11 (a) $\rho = 0.286$ (b) $r = 0.493$

12 (a) $\rho = 0$, $r = 0.123$
   (b) $\rho = -0.271$, $r = -0.445$
   (c) $\rho = -1$, $r = -0.987$

Intermediate

1 $\rho = -0.280$, $r = -0.260$
The small negative correlation makes it difficult to state categorically that marks go down as the temperature goes up. Only one set of results.

2 (a) $y = 3.2x$, $x = 0.0385 + 0.308y$
(b) $y = 14 - 1.25x$, $x = 10.9 - 0.763y$
(c) $y = 1.54 + 1.66x$,
  $x = -0.763 + 0.590y$
(d) $y = 11.7 - 0.366x$,
  $x = 30.6 - 2.45y$

3 (a) (i) $y = 0.5 + 1.5x$,
    $x = -0.143 + 0.643y$
  (ii) $y = 0.5 + 1.5x$,
    $x = -0.143 + 0.643y$
(b) (i) $y = 18.1 - 1.15x$,
    $x = 14.9 - 0.789y$,
  (ii) $y = 18.1 - 1.15x$,
    $x = 14.9 - 0.789y$

4 (a) $y = 0.3x + 0.4$
(b) $y = -0.5x + 0.85$
(c) $y = -2x + 0.5$
(d) $y = 0.2x + 2.4$
(e) $y = 1.25x$

5 $y = 2.9x - 4.9$

6 (b) $y = -18.0 + 0.556x$,
  $x = 35.8 + 1.65y$
(c) (i) $y = 20.9$
  (ii) $x = 85.3$ by calculation

7 (b) $y = 8.01 - 0.902x$,
  $x = 8.68 - 1.07y$
(c) (i) $y = 3.95$
  (ii) $x = 5.47$ by calculation

8 (a) $\rho = 0.1$, $r = 0.214$
(b) (i) Very low, positive
  (ii) P_1 improves, P_2 inconsistent but has more experience?

9 (a) (i) 0.464   (ii) 0.439
(b) Some, but not much, positive correlation
(c) 190.8 km/h

10 (b) 0.893
(c) $P = 420.3 + 0.880A$
(d) From equation: £454 000

* 11 (a) (i) $\Sigma x = 129$, $\Sigma x^2 = 2727$,
    $\Sigma y = 64$, $\Sigma y^2 = 628$,
    $\Sigma xy = 800$, $\bar{x} = 16.125$, $\bar{y} = 8$
  (ii) $S_{xy} = -29$, $S_x = 8.99$,
    $S_y = 3.81$
 *(iii)

Rank 1	1	2	3	4	5	6	7	8
Rank 2	7	8	5	3	6	4	2	1
$\lvert d \rvert$	6	6	2	1	1	2	5	7
d^2	36	36	4	1	1	4	25	49

$$\rho = 1 - \frac{6 \times 156}{8 \times 63} = 1 - 1.8571$$
$$= -0.857,$$
a high negative correlation
(b) (i) $\Sigma x = 34.9$, $\Sigma x^2 = 161.61$,
    $\Sigma y = 6.48$, $\Sigma y^2 = 5.3916$,
    $\Sigma xy = 28.27$, $\bar{x} = 4.3625$,
    $\bar{y} = 0.81$
  (ii) $S_{xy} = 0.000\,125$, $S_x = 1.0816$,
    $S_y = 0.134$
  (ii) $\rho = 0$

12 (a) (i) $y = 0.282 + 0.706x$,
    $x = 0.0324 + 0.130y$
  (ii) $y = 0.326$, $x = 0.0721$
(b) (i) $y = 2.85 - 0.0186x$,
    $x = 125 - 36.5y$
  (ii) $y = 1.55$, $x = 66.9$

13 (b) $F = -4.11 + 1.107M$; 60 to the nearest integer

14 (a) $\rho = 0.988$ (b) $r = 0.989$

*15

Rank 1	3	6	4	8	6	2	1	6
Rank 2	3	3	7	6	5	1	3	8
$\lvert d \rvert$	0	3	3	2	1	1	2	2
d^2	0	9	9	4	1	1	4	4

$$\rho = 1 - \frac{6 \times 32}{8 \times 63} = 0.619,$$
fairly high positive correlation

Advanced

1 (a) $\bar{x} = 4.2$, $\bar{y} = 3.25$
(b) $y = -0.651 + 0.929x$

(c) line 4.25 kg, calculation 4.36 kg
(d) Mass remains proportional to volume of Giantgro supplied.

2 (b) (56.9, 58.4)
(c) $y = 3.92 + 0.957x$
(d) 46
(e) Find the equation of the regression line of x on y.

*3 (a) If $\Sigma xy = 123$, then
$2 + 5A + 4A^2 + 70 = 123$
so $4A^2 + 5A - 51 = 0$,
i.e. $(4A + 17)(A - 3) = 0$
Hence $A = -\frac{17}{3}$ or 3, so $A = 3$
(b) $\Sigma x = 1 + 3 + 6 + 10 = 20$,
$\Sigma x^2 = 146, \Sigma xy = 123$
$\Sigma y = 2 + 5 + 6 + 7 = 20$,
$\Sigma y^2 = 114$
Hence line of y on x is
$$y - \frac{20}{4} = \frac{4 \times 123 - 20 \times 20}{4 \times 146 - 20 \times 20} \times \left(x - \frac{20}{4}\right)$$
so $y - 5 = \frac{1}{5}(x - 5)$ or $2y = x + 5$

4 (a) $r = 0.842$
(c) $L = 22.7 + 1.04S$,
$S = -12.2 + 0.684L$
(d) 37

5 (a) (i) $T = 4.13 + 0.265S$
(ii) $r = 0.309$
(b) $d = 0.373$

6 (a) Only ranks given, or ranks given as letters not numbers
(c) $r = -0.983$

7 (a) $P = 2, Q = 2$
(b) (i) -0.4 (ii) -0.4

8 (a) $\rho = 0.657$
(b) $O_1, O_3, O_5, O_2, O_6, O_4$ or 2, 3, 4, 1, 6, 5 or 2, 1, 6, 3, 4, 5, etc.

9 (a) $y = 3.41 + 0.679x$
(b) 10.8
(c) $x = -0.583 + 1.06y$; 11.2 seconds

10 (b) $y = -28.5 + 2.25x$ (c) 29.1

11 (b) $y = 0.246 + 1.09x$
(c) Ages of all over 17 overestimated.
(d) Peter tends to underestimate ages.
(e) $r = 0.960$ Strong positive linear relationship between the two estimates.
(f) $r = 0.960$
(g) Regression appropriate in (b) since estimates depend on actual age not vice versa. Correlation appropriate in (e) as their estimates do not depend on each other.

Revision

1 (a) $\Sigma x = 55, \Sigma y = 132, \Sigma x^2 = 385$,
$\Sigma y^2 = 1862, \Sigma xy = 629$,
$\bar{x} = 5.5, \bar{y} = 13.2$
(b) $y = 19.7 - 1.18x$,
$x = 16.2 - 0.811y$
(c) 8.25, -9.72, 12.0

2 (b) M(12.5, 6.5)
(c) (i) $y = 5.4$ (ii) $x = 15.3$

3 (a) $y = 20.7 - 0.583x$,
$x = 33.3 - 1.58y$
(b) $y = 34.4 - 1.52x$,
$x = 22.3 - 0.645y$

4 $x = 16.1 - 0.206y$; $x = 3.49$

5 (a) $y = -2 + 1.4x$
(b) $y = -3.7 + 0.7x$

6 $\rho = 0.657, r = 0.669$; a fair degree of positive correlation between J_1 and J_2

7 $\rho = 0.0714$. Wealth level of shoppers; other competing stores; weather

*8 $\Sigma x = 162.8, \Sigma y = 23.8$,
$\Sigma x^2 = 2662.74, \Sigma y^2 = 62.72$,
$\Sigma xy = 382.38, \bar{x} = 16.28, \bar{y} = 2.38$

(a) $r = \dfrac{10 \times 382.38 - 162.8 \times 23.8}{\sqrt{(26627.4 - 162.8^2)}\sqrt{(627.2 - 23.8^2)}}$

$= \dfrac{-50.84}{\sqrt{123.56}\sqrt{60.76}} = -0.587$

(b) y on x:

$y - 2.38 = \dfrac{-50.84}{123.56}(x - 16.28),$

i.e. $y = 9.08 - 0.411x$

x on y:

$x - 16.28 = \dfrac{-50.84}{60.76}(y - 2.38),$

i.e. $x = 18.3 - 0.837y$

9 $\rho = 0.818$; $T_2 = 6.09 + 0.721x$; $22\,°C$

10 $\rho = 0.952$; almost!

11 $\rho = 0.771$

12 (a) -0.6 (b) -0.722

13 (a) $r = 0.759$
(b) $r = -0.0917$
(c) $r = -1.0011 \approx -1$

14 (a) (i) $S_{xy} = 9.75$, $S_x^2 = 9.14$,
$S_y^2 = 10.9$
(ii) $y = -1.92 + 1.07x$,
$x = 2.13 + 0.893y$
(b) (i) $S_{xy} = -24.0$, $S_x^2 = 11.9$,
$S_y^2 = 49.3$
(ii) $y = 44.1 - 2.02x$,
$x = 21.7 - 0.487y$
(c) (i) $S_{xy} = 9.03$, $S_x^2 = 14.04$,
$S_y^2 = 5.94$
(ii) $y = 0.741 + 0.643x$,
$x = -0.964 + 1.52y$

Chapter 6 Sampling and estimation

6.1

Basic

2 (a) N(40, 8) (b) N(50, 9)
    (c) N(90, 17) (d) N(−10, 17)
    (e) N(260, 164) (f) N(0, 344)
    (g) N(−30, 41) (h) N(18, $\frac{17}{25}$)
    (i) N(32.5, 2.5625) (j) N(−110, 89)

3 (a) N(100, 1.6)
    (i) 0.2145 (iii) 0.7285
    (b) N(50, 1.5) (i) 0.5852 (ii) 0.1103
    (c) N(2, 0.0036)
    (i) 0.5897 (ii) 0.9938
    (d) N(1000, 900)
    (i) 0.0287 (ii) 0.8664

4 (a) 4.8, 0.44 (b) 0.032, 7.3 × 10⁻⁶

5 $\hat{\mu} = 1090$, $\hat{\sigma}^2 = 7377$

* 6 $\mu = \frac{1398}{28} = 49.9$

$$\hat{\sigma}^2 = \frac{28}{27}\left(\frac{70704}{28} - \left(\frac{1398}{28}\right)^2\right)$$

$$= 33.5$$

7 $\mu = 1958$, $\hat{\sigma}^2 = 9447$

8 $\mu = 2.5$, $\hat{\sigma}^2 = 0.72$

9 (a) 16, 17.5 (b) 4, 1.2
    (c) 2.3, 0.4 (d) $\frac{17}{3}$, 2
    (e) 50, 163 (f) 5.45, 0.943
    (g) 15.5, 12.1 (h) 4.7, 3.57
    (i) 8.1, 5.88 (j) 4.75, 0.558

10 (a) 10.8, 34.2 (b) 4.25, 5.07
    (c) 2.73, 0.008 51 (d) 5.5, 3.71
    (e) 21.5, 4.72

* 11 $\Sigma x = 2417$, $\Sigma x^2 = 486\,883$, so
    estimates are: mean $= \dfrac{2417}{12} = 201$ ml,

$$\text{variance} = \frac{12}{11}\left\{\frac{486\,883}{12} - \left(\frac{2417}{12}\right)^2\right\}$$

$$= 5.36$$

12 31.28, 0.000 369

13 78 g, 1.71 g, 0.0718

14 (a) 0.0668 (b) 0.8185

15 (a) 0.04 (b) 0.0333
    (c) 0.04 (d) 0.025

16 (a) (i) 0.9809 (ii) 0.9890 (iii) 0.5923
    (b) (i) 0.9076 (ii) 0.2827
    (c) (i) 0.8866 (ii) 0.0098

17 (a) 0.0092 (b) 0.6812

18 (a) (i) 3.182 (ii) 2.764 (iii) 0.816
    (iv) 1.943
    (b) (i) −1.886 (ii) −4.604
    (iii) −0.727 (iv) −14.09
    (c) (i) 3.747 (ii) 2.306 (iii) 2.015
    (d) (i) 2.353 (ii) 2.365 (iii) 1.372

Intermediate

1 (a) N(100, 25/n); 5
    (b) N(2, 0.16/n); 16
    (c) N(15.6, 3.5/n); 20
    (d) N(10, 0.64/n); 25
    (e) N(20, 4/n); 10

*2 (a) $B(40, 0.65) = N(26, 9.1)$, and for a sample size 30 this is

$$\bar{X} = N\left(26, \frac{9.1}{30}\right).$$

(i) $p(X \geq 28) = 1 - \Phi\left(\dfrac{1.5}{\sqrt{\dfrac{9.1}{30}}}\right)$

$= 1 - \Phi(2.724)$

$= 0.0033$

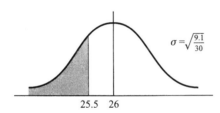

$\sigma = \sqrt{\dfrac{9.1}{30}}$

26   27.5

(ii) $p(X \leq 25) = 1 - \Phi\left(\dfrac{0.5}{\sqrt{\dfrac{9.1}{30}}}\right)$

$= 1 - \Phi(0.908)$

$= 0.1821$

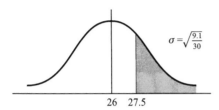

$\sigma = \sqrt{\dfrac{9.1}{30}}$

25.5   26

(b) (i) 0.9371   (ii) 0.6321
(c) (i) 0.5434   (ii) 0.4148
(d) 0.0855
(e) 0.9332

3

x	7.5	10	12.5
f	2	2	2

$\mu = 10$, $\sigma^2 = 16.7$, $E(X) = 10$,

$\text{Var}(\bar{X}) = 4.17 = \dfrac{\sigma^2(N - n)}{n(N - 1)}$

4 (a) Borage 75.7, 3.47; nasturtium 46.0, 1.90; cowslip 148, 5.54
(b) Second nasturtium sample, variance = 5.51

5 (a) (i) 0.0549, (ii) 0.02
(b) Same results as for (a)

6 (a) 0.0501   (b) 0.632   (c) $k = 41$

*7 $B(0.3 + 0.7)^{60} = N(42, 12.6)$

$p(X > 40) = \Phi\left(\dfrac{1.5}{\sqrt{12.6}}\right) = 0.6639$

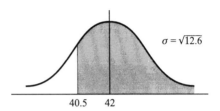

$\sigma = \sqrt{12.6}$

40.5   42

$p(X = 45) = \Phi\left(\dfrac{3.5}{\sqrt{12.6}}\right) - \Phi\left(\dfrac{2.5}{\sqrt{12.6}}\right)$

$= 0.0788$

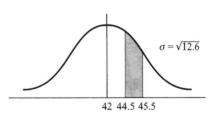

$\sigma = \sqrt{12.6}$

42 44.5 45.5

8 (a) $R = N(0.05, 0.000\,593\,8)$
(b) (i) 0.6779   (ii) 0.1190
(c) 0.1781

9 $A = 6$, $B = 5$; 1.97

10 (a) 1992   (b) 2819

11 (a) 3.7   (b) 0.03   (c) 0.0333

12 (a) 0.1068   (b) 0.2036

13 $\frac{5}{72} = 0.0694$, £14.62

14 0.289 or 28.9%

15 120

16 (a) 0.9   (b) 0.9975   (c) 0.975
(d) 0.025   (e) 0.999

17 (a) 0.0475   (b) 0.045   (c) 0.895
     (d) 0.899   (e) 0.985

18 85.6 ± 1.74 km h^{-1}

Advanced

*1 For one rod, N(0.44, 0.01); for one
bearing, N(0.47, 0.0144)

(i) $\text{p}(d > 0.465 \text{ cm}) = 1 - \Phi\left(\dfrac{0.025}{0.1}\right)$

$$= 0.4013$$

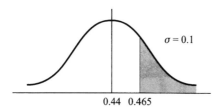

(ii) For 50 rods, N(22, 0.5) so

$$\text{p(total} < 20.5) = 1 - \Phi\left(\dfrac{1.5}{\sqrt{0.5}}\right)$$

$$= 0.017$$

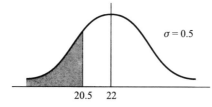

(iii) $\text{p}(d < 0.445) = 1 - \Phi\left(\dfrac{0.025}{0.12}\right)$

$$= 0.4176$$

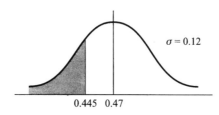

(iv) $R - B = \text{N}(-0.03, 0.0244)$,

$$\text{p}(R - B) < 0) = 1 - \Phi\left(\dfrac{0.03}{\sqrt{0.0244}}\right)$$

$$= 0.4239$$

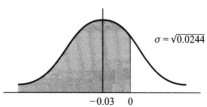

(b) For 100 rods, $\bar{D} = \text{N}(0.44, 0.0001)$

$$\text{p}(0.42 < D < 0.45)$$

$$= \Phi\left(\dfrac{0.01}{\sqrt{0.0001}}\right)$$

$$+ \Phi\left(\dfrac{0.02}{\sqrt{0.0001}}\right) - 1$$

$$= 0.8185$$

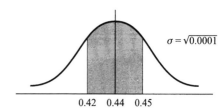

2 (a) (i)

Mean	3	4	5	6	7	8	9	10	11	12
Frequency	1	3	6	10	12	12	10	6	3	1

(ii)

Mean	6	7	8	9	
Frequency	6	6	6	6	$\mu = 7.5$,

$$\sigma^2 = 11.25$$

3 (a) (i) 0.006   (ii) 0.2182   (iii) 0.0851
      (iv) 0.1930
     (b) £2

4 (a) $\frac{3}{14}$ (b) $\text{E}(X) = \frac{10}{7}$, $\text{Var}(X) = \frac{46}{245}$
     (c) 0.2139

5 (b) $\text{E}(X) = \frac{23}{24}$, $\text{Var}(X) = \frac{341}{1728}$
     (c) 0.7464

6 (a) 0.1919   (b) 3.28, 1.65

7 $\frac{2}{15}, \frac{3}{26}$, 123

8 (a) (i) Mean 2 3 4 5 6
          f 1 2 3 2 1

    (ii) Mean 3 4 5
         f 2 2 2

   (b) Mean = 4, variance = $\frac{8}{3}$

Revision

1 (a) 8.24, 7.30   (b) 5.68, 0.155
   (c) 16.4, 40.5

2 (a) 0.1029   (b) 0.3027

3 (a) 50   (b) (i) 0.1831   (ii) 0.1958

4 (a) 0.0938   (b) 0.0896

5 (a) 0.0363   (b) 0.2041

6 (a) 2.158
   (b) (i) 0.000 306   (ii) 0.000 306
   (c) 0.000 322

7 (a) 5.59, 0.122   (b) 0.124

*8 (a) 36   (b) 25

   *(c) $\bar{X} = N\left(10, \dfrac{\sigma^2}{10}\right),$

   $p(\bar{X} > 10.1) = 1 - \Phi\left(\dfrac{0.1}{\sigma/\sqrt{10}}\right)$

   $= 0.4164$

   $\Phi\left(\dfrac{0.1}{\sigma/\sqrt{10}}\right) = 1 - 0.4164$

   $= 0.5836 = \Phi(0.211),$

   hence $\sigma^2 = 2.25$

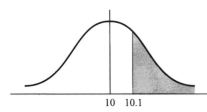

   10 10.1

   (d) 3.50

9 (a) 146   (b) 2.08   (c) 1020

10 (a) $k = \frac{2}{5}$ (b) $\frac{7}{5}, \frac{31}{150}$ (c) 50

11 (a) 0.161 or 16.1%
    (b) £6 255 000
    (c) (i) 0.1345   (ii) 0.8918

12 61.7 ≈ 62

6.2

Basic

1 (a) (i) 7   (ii) 7 ± 1.10
   (b) (i) 11.8   (ii) 11.8 ± 0.987
   (c) (i) 0.58   (ii) 0.58 ± 0.147

2 (a) (i) 152 ± 0.576
       (ii) 152 ± 0.438
   (b) (i) 25.4 ± 0.141
       (ii) 25.4 ± 0.107
   (c) (i) 156.3 ± 0.178
       (ii) 156.3 ± 0.136
   (d) (i) 241 ± 4.72
       (ii) 241 ± 3.59

3 (a) (i) 50 ± 1.18 (ii) 50 ± 1.55
   (b) (i) 90 ± 0.784 (ii) 90 ± 0.930
   (c) (i) 40 ± 1.47 (ii) 40 ± 1.93

*4 (a) CI $= 25.2 \pm 1.645\sqrt{\dfrac{0.4}{35}}$

       $= 25.2 \pm 0.176$ kg

   (b) CI $= 25.2 \pm 2.326\sqrt{\dfrac{0.4}{35}}$

       $= 25.2 \pm 0.249$

5 8 minutes 12 seconds $\pm$ 17.7 seconds

6 2.02 ± 0.0544 kg

7 174 g; 174 ± 18.5 g

8 (a) (i) 0.25 ± 0.134 (ii) 0.25 ± 0.176
   (b) (i) 0.2 ± 0.0658 (ii) 0.2 ± 0.0930
   (c) (i) 0.15 ± 0.0990 (ii) 0.15 ± 0.117
   (d) (i) 0.2 ± 0.101 (ii) 0.2 ± 0.133
   (e) (i) 0.14 ± 0.0571 (ii) 0.14 ± 0.0807

9 0.1 ± 0.0698; 302–1698

10 (a) 0.1 ± 0.0588 (b) 0.6 ± 0.0681
       (c) 0.25 ± 0.144 (d) 0.85 ± 0.117
       (e) 0.95 ± 0.0334

11 (a) $(27.8, \infty)$, $(-\infty, 32.2)$
       (b) $(2.51, \infty)$, $(-\infty, 3.4)$
       (c) $(3.76, \infty)$, $(-\infty, 5.04)$
       (d) $(14.1, \infty)$, $(-\infty, 16.5)$

12 $(0, 0.605)$

13 (a) 156 ± 1.75, 156 ± 2.69
       (b) 36.9 ± 1.32, 36.9 ± 2.09
       (c) 76.7 ± 2.82, 76.7 ± 4.71
       (d) 749 ± 1.81, 749 ± 2.74
       (e) 75 ± 0.778, 75 ± 1.14
       (f) 514 ± 3.74, 514 ± 6.57

Intermediate

1 0.36 ± 0.0210; 7 620 000, 6 780 000

2 0.43 ± 0.097; 4996, 3157

3   (a) 79.3 hours, 414
       (c) 416
       (d) 79.3 ± 3.71 hours

4   (a) 76.3 kg, 14.5
       (b) 14.8
       (c) 76.3 ± 1.08 kg

5   (a) $\bar{d} = 0.5\,\text{mm}$, $s^2 = 0.000\,19$
       (b) $\hat{\sigma}^2 = 0.0002$
       (c) 0.5 ± 0.00736; 0.2195

6 0.00714 ± 0.00624; 47, 3

*7 $\bar{W} = \frac{2050}{25} = 82$ kg
       $s^2 = \frac{168\,402}{25} - \left(\frac{2050}{25}\right)^2 = 12.08 \approx 12.1$
       $\hat{\sigma}^2 = \frac{25}{24}\left\{\frac{168\,402}{25} - 82^2\right\} = 12.58333 \approx 12.6$
       For the sample N(82, 12.08)

       $p(\bar{W} > 90) = 1 - \Phi\left(\dfrac{8}{12.08}\right)$

       $\qquad\qquad = 1 - \Phi(2.302)$

       $\qquad\qquad = 1 - 0.9894$

       $\qquad\qquad = 0.0106$

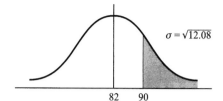

$\sigma = \sqrt{12.08}$

82   90

$$CI = 82 \pm 1.645\sqrt{\frac{12.583\dots}{25}}$$
$$= 82 \pm 1.1671 \approx 82 \pm 1.17$$

8   (a) 0.38   (b) 0.38 ± 0.0885

9   (a) 2366   (b) 310
       (c) 983   (d) 86

10 (a) 11.7 hours   (b) 0.206 hours
       (c) 0.0441   (d) 11.7 ± 0.0677 hours

11 13

12 (a) 41   (b) 19   (c) 199
       (d) 16   (e) 145

13 7

14 $(0, 0.101)$

15 $(0, 10.5)$

16 1.84 ± 0.00773

17 496 ± 2.88

Advanced

1   (a) 0.0503   (b) 0.0468
       (c) 0.0468 ± 0.0227
       (d) 245 should be tested

2   313; 500, 228; 252–411

3   (a) 0.2 ± 0.0931
       (b) $(0.127, 1)$
       (c) 53, 64
       (d) (a), (b) unchanged, (c) 42, 50

4   (a) 8   (b) 28   (c) 14
       (d) 3.50 ± 0.980

5 (a) 255 g, 1.72 g   (b) 3.07 g
(c) 255 ± 0.744 g

6 (a) 50   (b) 6.69   (c) 6.83
(d) 24.7 ± 0.759

7 (a) $\frac{2}{15}$ (b) 0.133 ± 0.0122

*8 $\bar{x} = \dfrac{285.08}{20} = 14.254,$

$\hat{\sigma}^2 = \left\{ \dfrac{4063.5338}{20} - (14.254)^2 \right\} \times \dfrac{20}{19}$

$= 0.000183$
CI is 14.254 ± 0.00593

9 1.13, 0.0603; 1.13 ± 0.06; large sample,
or Central Limit Theorem applies

Revision

1 (a) 5.2 ± 0.134   (b) 48 ± 1.37
(c) 8.6 ± 0.431   (d) 126 ± 3.86

2 (a) 5.71 ± 1.33   (b) 12 ± 1.39
(c) 23 ± 1.73   (d) 6.42 ± 0.575

3 (a) 4.99; 9.60 ± 1.29
(b) 20.7; 4.27 ± 1.33
(c) 8.86; 123 ± 1.26

4 (a) (i) 3.8, 1.76   (ii) 1.89
(iii) 3.8 ± 0.583
(b) (i) 7.5, 3.05   (ii) 3.39
(iii) 7.5 ± 0.958
(c) (i) 4.48, 0.324   (ii) 0.364
(iii) 4.48 ± 0.331

5 0.0293

6 (0, 0.472)

7 48 ± 0.791, 48 ± 0.942

8 (a) 3.99, 0.0746   (b) 3.99 ± 0.120

9 0.131; 0.136 ± 0.0641

10 (a) 82.8 kg   (b) 24.8
(c) 82.8 ± 1.73

11 (a) 0.25 ± 0.0582   (b) 0.3 ± 0.0688
(c) 0.65 ± 0.101   (d) 0.5 ± 0.0823

*12 $\Sigma x = 8791$, $\Sigma x^2 = 8\,092\,925$, $n = 10$,
so $\bar{x} = 879.1$

$\hat{\sigma}^2 = \left(\dfrac{8\,092\,925}{10} - (879.1^2) \right)$

$\times \dfrac{10}{9} = 40\,528.544.$

so CI is $879.1 \pm 1.96\sqrt{\dfrac{40\,528.544}{10}}$,
i.e. 879 ± 125.
Sample proportion
$= 1 - \frac{36}{120} = \frac{84}{120} = 0.7$
So, CI is
$0.7 \pm 1.96\sqrt{\dfrac{0.7 \times 0.3}{120}}$, i.e. 0.7 ± 0.0820

Chapter 7 Hypothesis testing

7.1

Basic

1 Abbreviations: A means 'accept H_0', R means 'reject H_0'

(a) $Z = 1.33$	(i) A	(ii) A	(iii) R
(b) $Z = -2$	(i) A	(ii) R	(iii) R
(c) $Z = -1.43$	(i) A	(ii) A	(iii) A
(d) $Z = 2.47$	(i) A	(ii) R	(iii) R
(e) $Z = 1.75$	(i) A	(ii) A	(iii) R
(f) $Z = 1.74$	(i) A	(ii) R	(iii) R
(g) $Z = 1.6$	(i) A	(ii) A	(iii) A
(h) $Z = -2.59$	(i) R	(ii) R	(iii) R
(i) $Z = 2.07$	(i) A	(ii) R	(iii) R
(j) $Z = -1.76$	(i) A	(ii) R	(iii) R

2 $Z = 1.89$, $Z > 1.645$; reject H_0, average > 64%

3 $Z = 1.68$, $Z > 1.645$; reject H_0, average > 280 g

*4 H_0: $\mu = 85$ kg, H_1: $\mu < 85$ kg,
so $Z = \dfrac{66 - 85}{7.4} = -2.57$

At 5% level with a one-tailed test, $|Z| > 1.645$, so reject H_0; there is significant evidence at the 5% level that the average mass < 85 kg.

5 $Z = 2.29$, $Z < 2.575$; accept H_0, average = 0.22 cm

6 $Z = 0.866$, $Z < 1.282$; accept H_0, die not biased

7 $Z = -2$, $|Z| > 1.645$; reject H_0, average below 20

8 (a) $Z = -1.93$ (i) Accept (ii) Reject
(b) $Z = 1.54$ (i) A (ii) R
(c) $Z = -1.45$ (i) A (ii) R
(d) $Z = 1.49$ (i) A (ii) A
(e) $Z = 2.13$ (i) R (ii) R

9 $\hat{\sigma} = 0.403$, $Z = 2.22$, $Z > 2.045$; reject H_0, mean > 100 g

10 $t = -2.82$, $|t| > 2.718$; reject H_0, mean < 76 g

11 $t = 1.30$, $t < 1.833$; accept H_0, mean = 1.84 km

*12 H_0: $\mu = 85$ kg, H_1: $\mu < 85$ kg
At 2% significance level, with a one-tailed test,

estimate of population variance
$$= (0.93)^2 \left(\tfrac{32}{31}\right) = 0.8928$$
test statistic,
$$Z = \frac{5.85 - 6.2}{\sqrt{0.8928/32}} = -2.095$$

$|Z| > 2.054$, so reject H_0; there is significant evidence, at the 2% level, that the mean has fallen below 6.2 g.

Intermediate

1 $t = -2.14$, $|t| > 1.711$; reject H_0

*2 H_0: $\mu = 750$ ml, H_1: $\mu \neq 750$ ml. At 5% significance level with a two-tailed test,

test statistic, $Z = \dfrac{749.7 - 750}{\sqrt{1.21/50}} = -1.93$

$|Z| < 1.96$; accept H_0, no change at this level

3 (a) $Z = 2.12$, $Z > 1.645$; reject H_0

(b) $Z = 2.24$, $Z > 1.645$; reject H_0

4 $Z = -1.73$, $|Z| > 1.645$; reject H_0

5 (a) $t = 1.352$, $t < 1.706$; accept H_0

(b) $t = 0.756$, $t < 2.093$; accept H_0

(c) $t = 1.93$, $t > 1.706$; reject H_0

(d) $Z = -1.88$, $|Z| > 1.282$; reject H_0

6 $Z = -2.59$, $|Z| > 1.282$; reject H_0

7 S_1: $t = -2.00$, $|t| > 1.383$; reject H_0,

S_2: $t = -0.801$, $|t| < 1.383$; accept H_0.

*8 $p_s = \frac{65}{200} = \frac{13}{40}$, H_0: $p(D) = \frac{1}{4}$,

H_1: $p(D) > \frac{1}{4}$

With one-tailed test at 2% significance level,

test statistic, $Z = \dfrac{\frac{13}{40} - \frac{1}{4} - \frac{1}{400}}{\sqrt{\left(\frac{\frac{1}{4} \times \frac{3}{4}}{200}\right)}} = 2.37$

$Z > 2.054$, so reject H_0; there is significant evidence at the 2% level that the spinner is biased towards the sector D.

9 $Z = 1.91$, $Z > 1.645$; reject H_0, rate has increased.

10 $t = -2.30$, $|t| < 2.539$; accept H_0

11 $Z = -2.055$, $|Z| < 2.326$; accept H_0, mean $= 2.76$

12 $Z = -1.34$, $|Z| > 1.282$ but $|Z| < 1.645$

13 $T = 12.2$ seconds, $t = -1.12$, $|t| = 1.12$, $t < 2.821$; accept H_0

14 $t = 1.58$, $t < 1.833$; no, there is not significant evidence that it has risen.

15 $Z = -1.167$, $|Z| < 1.645$; no, mean consumption has not gone down.

Advanced

1 $\bar{x} = 501.7$, $s^2 = 42.21$, $\hat{\sigma}^2 = 43.67$,

$Z = 1.409 < 1.645$; accept H_0, mean is 500 g

2 (a) 0.0094

(b) $Z = -2.34$, $|Z| > 2.054$, so reject H_0, $\bar{S}_2 < \bar{S}_1$

3 (a) 0.3 ± 0.0759

(b) $Z = 1.269 < 1.282$; accept H_0

*4 $p_s = \frac{98}{500}$, H_0: $p(6) = \frac{1}{6}$, H_1: $p(6) > \frac{1}{6}$

With one-tailed test,

test statistic, $Z = \dfrac{\frac{98}{500} - \frac{1}{6} - \frac{1}{1000}}{\sqrt{\left(\frac{\frac{1}{6} \times \frac{5}{6}}{500}\right)}} = 1.7$

$|Z| > 1.645$, so H_0 is rejected; there is significant evidence at the 5% level to show the die is biased. The 99% confidence interval is

$\frac{90}{500} \pm 2.575 \sqrt{\dfrac{\frac{90}{500} \times \frac{410}{500}}{500}} = 0.18 \pm 0.0442$

Again, H_0: $p(6) = \frac{1}{6}$, H_1: $p(6) > \frac{1}{6}$

With one-tailed test,

test statistic, $Z = \dfrac{\frac{90}{500} - \frac{1}{6} - \frac{1}{1000}}{\sqrt{\left(\frac{\frac{1}{6} \times \frac{5}{6}}{500}\right)}} = 0.74$

$Z < 1.645$, so accept H_0; this die is not biased at the 5% level.

5 $Z = -0.210$, $|Z| < 1.645$, accept H_0;

$Z = 0.109$, $Z < 1.96$, accept H_0

6 $Z = -2.04$, $|Z| < 2.054$, (just) accept H_0; 8.00 minutes

7 $Z = 2.858$, $Z > 2.326$, reject H_0; H_1 is justified.

8 $Z = -1.167$, $|Z| < 1.645$, accept H_0; consumption did not fall by more than 2%

$Z = 0.885$, $|Z| < 1.645$, accept H_0; average consumption has not changed

9 (a) $Z = -1.22$, $Z_c = -1.6449$, accept H_0

(b) Range 33 or 11 SDs. Almost impossible for normal distribution.

(c) (i) Unaffected – large sample, mean is approx normal.

(ii) Unaffected – Z would be numerically less than 1.22.

Revision

1 (a) $Z = 1.155$, $Z < 1.645$; accept H_0, $\mu = 65$

(b) $Z = -2.271$, $|Z| > 1.96$; reject H_0, $\mu \neq 260$

(c) $Z = 1.167 < 1.282$; accept H_0, $\mu = 0.45$

(d) $Z = 2.593 > 2.575$; reject H_0, $\mu \neq 12.7$

(e) $Z = -1.656$, $|Z| > 1.645$; reject H_0, $\mu < 30.6$

2 (a) $Z = -1.832$, $|Z| < 1.96$; accept H_0, $\mu = 22.3$

(b) $Z = -2.47$, $|Z| > 2.326$; reject H_0, $\mu < 5.24$

(c) $Z = 2.120 > 2.054$; reject H_0, $\mu > 0.04$

(d) $Z = -2.050$, $|Z| < 2.17$; accept H_0, $\mu = 7.92$

(e) $Z = 1.850 > 1.751$; reject H_0, $\mu > 48.9$

3 $Z = -1.82$, $|Z| > 1.645$; reject H_0, $\mu < 250$

4 $Z = 1.421$, $Z < 1.645$; accept H_0, proportion $= 1\%$

*5 H_0: $D = 1.84\,\text{cm}$, H_1: $D > 1.84\,\text{cm}$

With a one-tailed test at 5% level,

test statistic, $Z = \dfrac{1.86 - 1.84}{\sqrt{\left(\dfrac{0.0072}{50}\right)}} = 1.67$

$Z > 1.645$, so H_0 is rejected; there is significant evidence at the 5% level that the diameter has increased and the cutting surfaces should be replaced.

6 $Z = 1.601$, $Z > 1.282$; reject H_0, level has increased

7 $Z = -2.438$, $|Z| > 2.326$; reject H_0, $p < 0.6$

8 $Z = -2.486$, $|Z| > 2.326$; reject H_0, $\bar{X}_2 > \bar{X}_1$

9 (a) $Z = -2.235$, $|Z| > 1.96$; reject H_0, $\mu \neq 156$

(b) $Z = 1.874$, $Z > 1.645$; reject H_0, $\mu > 156$

10 $Z = 1.914$, $Z < 1.96$; accept H_0, no significant difference in the means $Z = -1.262$, $|Z| < 1.282$; accept H_0, $\mu_1 = 21\,\text{oz}$

11 $Z = 1.424$, $Z < 1.645$; accept H_0, $\mu_1 = \mu_2$

12 $Z = -1.722$, $|Z| > 1.645$; reject H_0, claim too high $Z = -1.964$, $|Z| > 1,645$; reject H_0, again claim too high

13 $\hat{\sigma}^2 = 10.2$, $t_5\,5\% = 2.015$, $t = 1.764$, $|t| < 2.015$, so accept H_0

14 $t_{11}\,5\% = 1.796$, $t = -2.31$, $|t| > 1.796$, so reject H_0

7.2

Basic

1 (a) 0.017   (b) 51.2

2 0.2228

3 0.709, 0.0320

4 $X > 70.2$, 0.158

5 (a) (i) 0.859 (ii) 0.75 (iii) 0.438

(b) (i) 0.117 (ii) 0.237 (iii) 0.616

6 0.0143

7 0.3222

8 (a) $X > 52.403$
(b) (i) 0.8318 (ii) 0.3414 (iii) 0.0377

*9 H_0: p(heads) $= 0.5$, H_1: p(heads) > 0.5
Binomial is $X = B(\frac{1}{2} + \frac{1}{2})^{50}$
$\qquad = N(25, 12.5)$
H_0 is accepted if number of heads < 30.
Significance level $=$ p(Type I error)
$$= 1 - \Phi\left(\frac{4.5}{\sqrt{12.5}}\right)$$
$$= 1 - \Phi(1.273)$$
$$= 0.1014,$$
i.e. 10.14%

10 (a) 0.2018   (b) 0.6561

11 0.2458

12 (a) 0.46   (b) 0.497

Intermediate

1 (a) (i) 0.583 (ii) 0.853 (iii) 0.313
(b) (i) 0.673 (ii) 0.483 (iii) 0.265
(c) (i) 0.027 (ii) 0.064 (iii) 0.343
(d) (i) 0.669 (ii) 0.794 (iii) 0.965

2 (a) 0.663   (b) 0.188

3 (a) 0.711   (b) 0.244

4 (a) 40.7%   (b) 0.924   (c) 0.728

5 (a) 0.75   (b) $z = 0.472$

6 (a) 0.0202   (b) 0.0107; $k = 112$

7 (a) 0.1147   (b) 11

8 (a) 0.603   (b) 0.590   (c) 0.0549

9 $M = 505$ g

10 (a) 0.0207, 2.07%   (b) 96.9

11 (a) 0.0955, 9.55%   (b) 0.8302

12 (a) 0.405   (b) 83.3%

13 (a) 0.945   (b) 0.571

*14 (a) p(H_0 true) $= \frac{1}{6}\int_{1/2}^{2}(2x + 1)\,dx$
$\qquad = \frac{1}{6}[x^2 + x]_{1/2}^{2}$
$\qquad = \frac{1}{6}(4 + 2 - \frac{1}{4} - \frac{1}{2})$
$\qquad = 0.875$
Hence p(Type I error) $= 1 - 0.875$
$\qquad\qquad = 0.125$
(b) p(H_1 true) $= \frac{1}{2}\int_{0}^{1/2} x\,dx$
$$= \frac{1}{2}\left[\frac{x^2}{2}\right]_{0}^{1/2} = \frac{1}{2}(\frac{1}{8}) = \frac{1}{16}$$
Hence p(Type II error)
$\qquad = \frac{15}{16} = 0.938$

15 (a) 0.927   (b) 0.890

Advanced

1 (a) (i) 76.6–83.4 (ii) 77.1–82.9
(iii) 77.9–82.1
(b) (i) 0.903 (ii) 0.946 (iii) 0.655

2 (a) 0.438   (b) 0.325, 32.5%

3 0.0333, 0.015

4 $\bar{X} < 6.88$; 0.0016, power $= 99.84\%$

5 (a) 0.775   (b) 0.3625 or 36.25%

7 (a) p(2) $+$ p(3) $+$ p(4)
$= {}^4C_2(\frac{8}{20})^2(\frac{12}{20})^2 + 4(\frac{8}{20})(\frac{12}{20})^3 + (\frac{12}{20})^4$
$= 0.3456 + 0.3456 + 0.1296$
$= 0.8208$
So p(Type I error) $= 1 - 0.8208$
$\qquad\qquad = 0.1792$
(b) p(Type II error)
$= 6(\frac{12}{20})^2(\frac{8}{20})^2 + 4(\frac{12}{20})(\frac{8}{20})^3 + (\frac{8}{20})^4$
$= 0.3456 + 0.1536 + 0.0256$
$= 0.5248$

(c) Power $= 1 - 0.5248 = 0.4752$ or 47.52%

8 0.554, 0.276

9 (i) $k = 300.4$
(ii) 0.033
(iii) Yes, reasonably so; possible errors of 5% and 3.3%
(iv) 0.97

Revision

2 (a) $\frac{11}{12}$ (b) $\frac{119}{120} = 0.992, 99.2\%$
3 (a) 3.02–3.38
(b) (i) 0.8364
(ii) 0.9499
(iii) 0.7785

4 (a) (i) 0.0359   (ii) 0.7881
(b) 0.9711

5 0.102

*6 $1 - \Phi\left(\dfrac{72 - k}{4/\sqrt{15}}\right) = 0.1,$

$$\Phi\left(\dfrac{72 - k}{4/\sqrt{15}}\right) = 0.9 = \Phi(1.282)$$

so $k = 72 - \dfrac{4 \times 1.282}{\sqrt{15}} = 70.7$

7 $k = 6.23$

8 0.1064, 0.7967

9 0.48

10 (a) 0.467   (b) 0.296

Chapter 8 χ^2 tests

Basic

1 (a) (i) 0.990   (ii) 0.050   (iii) 0.950
    (iv) 0.999   (v) 0.975
  (b) (i) 0.050   (ii) 0.010   (iii) 0.001
    (iv) 0.950   (v) 0.025

2 (a) (i) 11.34   (ii) 0.8312   (iii) 15.51
    (iv) 3.940   (v) 71.42
  (b) (i) 0.1026   (ii) 16.81   (iii) 0.8312
    (iv) 5.229   (v) 21.03

3 (a) 0.024   (b) 0.025   (c) 0.024
  (d) 0.015   (e) 0.040

4 (a) 0.010   (b) 0.025   (c) 0.001
  (d) 0.025   (e) 0.05

*5

Score	1	2	3	4	5	6
f_0	35	27	37	55	51	35
f_e	40	40	40	40	40	40
$O-E$	−5	−13	−3	15	11	5
$(O-E)^2$	25	169	9	225	121	25
$\dfrac{(O-E)^2}{E}$	0.625	4.225	0.225	5.625	3.025	0.625

χ^2_5 5% = 11.07, χ^2_{test} = 14.35, die is not fair

---

6 χ^2_3 5% = 7.815, χ^2_{test} = 7.18 < χ^2_3 5%, so die is fair at 5%
χ^2_3 10% = 6.251, so die is not fair at 10%

7 χ^2_1 10% = 2.706, χ^2_{test} = 2.501 (Yates' correction used)
Yes, it does agree with the National Survey.

8 (a) χ^2_5 1% = 15.09, χ^2_{test} = 15.21; no
  (b) χ^2_9 5% = 16.92, χ^2_{test} = 17.37; no
  (c) χ^2_3 10% = 6.251, χ^2_{test} = 4.16; yes
  (d) χ^2_6 5% = 12.59, χ^2_{test} = 8.12; yes
  (e) χ^2_5 1% = 15.09, χ^2_{test} = 19.04; no

9 (a) χ^2_4 10% = 7.779, χ^2_{test} = 8.293; no
  (b) χ^2_5 5% = 11.07, χ^2_{test} = 6.017; yes
  (c) χ^2_4 1% = 13.28, χ^2_{test} = 13.63; no
  (d) χ^2_3 10% = 6.251, χ^2_{test} = 6.128; yes
  (e) χ^2_5 2.5% = 12.83, χ^2_{test} = 13.27; no
  (f) χ^2_4 10% = 7.779, χ_{test} = 9.09; no

10 χ^2_2 1% = 9.21, χ^2_{test} = 11.43; no, X does not have a binomial distribution

11 (a) Expected frequencies

       6   14 χ^2_{test} = 2.232 < 3.841, independent
      12   28
  (b)   8   32 χ^2_{test} = 11.00 > 3, 841, dependent
      12   48
  (c) 15   15 χ^2_{test} = 3.857 > 3.841, (just) dependent
      35   35
  (d) 20.9   23.1 χ^2_{test} = 2.786, independent
      36.1   39.9
  (e) 37.35   52.65 χ^2_{test} = 0.676, independent
      45.65   64.35

Intermediate

1 χ_9^2 1% = 21.67, χ_{test}^2 = 23.86; numbers are not random

*2

	A	B	C	D	E
O	145	356	497	312	90
E	168	336	490	294	112
$O - E$	23	20	7	18	22

$$\Sigma \frac{(O-E)^2}{E} = 31.49 + 1.190 + 0.100$$
$$+ 1.102 + 4.321$$
$$= 9.862$$

χ_4^2 2.5% = 11.14, χ_{test}^2 = 9.862; no, they don't differ significantly at this level

3 χ_4^2 10% = 7.779, χ_{test}^2 = 6.422; yes, they do fit

4 χ_4^2 10% = 7.779, χ_{test}^2 = 13.521; yes, bias is shown

5 χ_3^2 5% = 7.815, χ_{test}^2 = 9.199; yes, they should change proportions

6 (a) χ_2^2 10% = 4.605, χ_{test}^2 = 8.382; no, they do not fit
(b) χ_2^2 5% = 5.991, χ_{test}^2 = 3.43; yes, they do fit
(c) χ_4^2 2.5% = 11.14, χ_{test}^2 = 12.4; no, they don't fit

7 (a) χ_5^2 1% = 15.09, χ_{test}^2 = 16.1; no, die not fair
(b) χ_1^2 1% = 6.635, χ_{test}^2 = 1.344; yes, it could have a binomial distribution

8 χ_4^2 10% = 7.779, χ_4^2 5% = 9.488
(a) χ_{test}^2 = 5.977   (b) χ_{test}^2 = 8.527

9 χ_6^2 2.5% = 14.45, χ_6^2 5% = 12.59, χ_{test}^2 = 13.597

10

	Po(3)	Po(4)
χ_{test}^2	9.764	90.167

χ_5^2 5% = 11.07, figures fit Po(3) but not Po(4)

11 χ_6^2 1% = 16.81, χ_{test}^2 = 13.208; yes, they could fit Po(5)

12 χ_4^2 10% = 7.779, χ_{test}^2 = 7.276; H_0 accepted, no significant improvement

*13

	5	~ 5
O	125	175
p(that score)	$\frac{1}{3}$	$\frac{2}{3}$
E	100	200

$$\Sigma\left\{\frac{(O-E)-\frac{1}{2}}{E}\right\}^2 = \frac{24.5^2}{100} + \frac{24.5^2}{200}$$

$$= 9.00375 \approx 9.00$$

χ_1^2 1% = 6.635,

χ_{test}^2 = 9.00, χ_{test}^2 > χ_1^2 1%
H_0 is rejected, there is significant evidence at this level that the die is biased towards a five.

14 χ_6^2 10% = 10.64, χ_{test}^2 = 7.76; yes, it could be described by N(4, 8)

Advanced

1 χ_5^2 1% = 15.09, χ_{test}^2 = 13.26; yes, they do fit the rule

2 μ = 21.68, $\hat{\sigma}^2$ = 16.5, χ_2^2 5% = 5.991, χ_{test}^2 = 6.885; no, they do not follow a normal distribution.

*3 $\mu = \frac{1}{100}(5 + 16 + 36 + 76 + 100 + 90$
$+ 63 + 48)$
$= \frac{444}{100} = 4.44$ hours

Expected hours:
$$e^{-4.44} \times 100 = 1.17959$$

$$e^{-4.44} \times 4.44 \times 100 = 5.23739$$

$$e^{-4.44} \times \frac{4.44^2}{2!} \times 100 = 11.62702$$

$$e^{-4.44} \times \frac{4.44^3}{3!} \times 100 = 17.20799$$

$$\vdots$$

$$e^{-4.44} \times \frac{4.44^8}{8!} \times 100 = 8.17273$$

0	1	2	3	4	5	6	7	8
O 4	5	8	12	19	22	15	9	6
E 1.179 59	5.237 39	11.627 02	17.207 99	19.100 87	16.961 57	12.551 56	7.961 27	8.172 73

Columns 0 and 1 must be combined to give an E value greater than 5; this gives

O	9
E	6.416 98

Hence

$$\chi^2_{\text{test}} = 1.0397 + 1.1314 + 1.5672$$
$$+ 0.000\,533 + 1.4967 + 0.4776$$
$$+ 0.013\,552\,6 + 0.5776$$
$$= 6.313\,366\,9$$

$\chi^2_7\ 10\% = 12.02$, $\chi^2_{\text{test}} = 6.313$

Yes, they do fit Po(4.44).

5 $\chi^2_1\ 1\% = 6.635$, $\chi^2_{\text{test}} = 14.69$; yes, there is a significant difference

6 $\chi^2_1\ 2.5\% = 5.024$, $\chi^2_{\text{test}} = 8.808$; yes, the choice depends on area

7 $\chi^2_2\ 5\% = 5.991$, $\chi^2_{\text{test}} = 6.015$; yes, the choice depends on gender

8 $\chi^2_2\ 10\% = 4.605$, $\chi^2_{\text{test}} = 3.887$; no difference

9 342.7, 117.7, 314.3; $\Sigma \dfrac{(f_o - f_e)^2}{f_e} = 14.8$

(with Yates correction, 14.22)
Use of machines by gender is significantly different.

Revision

1 (a) $\chi^2_4\ 5\% = 9.488$, $\chi^2_{\text{test}} = 7.71$; yes, it fits B(4, 0.6)

(b) $\chi^2_3\ 5\% = 7.815$, $\chi^2_{\text{test}} = 8.613$; no, it does not fit B(5, $\frac{1}{2}$)

(c) $\chi^2_3\ 10\% = 6.251$, $\chi^2_{\text{test}} = 7.245$; no, it does not fit B(4, 0.35)

2 $\chi^2_7\ 10\% = 12.02$, $\chi^2_{\text{test}} = 4.259$; yes, they do fit B(7, 0.52)

3 (a) $\chi^2_3\ 10\% = 6.251$, $\chi^2_{\text{test}} = 10.22$; no, reject H_0

(b) $\chi^2_5\ 5\% = 7.815$, $\chi^2_{\text{test}} = 4.357$; yes, accept H_0

(c) $\chi^2_4\ 2.5\% = 11.14$, $\chi^2_{\text{test}} = 13.15$; no, reject H_0

(d) $\chi^2_4\ 10\% = 6.251$, $\chi^2_{\text{test}} = 3.883$; yes, accept H_0

(e) $\chi^2_5\ 1\% = 15.09$, $\chi^2_{\text{test}} = 10.04$; yes, accept H_0

(f) $\chi^2_4\ 5\% = 9.488$, $\chi^2_{\text{test}} = 12.671$; no, reject H_0

4 (a) $\chi^2_3\ 2.5\% = 9.348$, $\chi^2_{\text{test}} = 9.714$; no, it does not fit Po(1.4)

(b) $\chi^2_5\ 10\% = 9.236$, $\chi^2_{\text{test}} = 9.795$; no, it does not fit Po(2)

5 $\chi^2_4\ 10\% = 7.779$, $\chi^2_{\text{test}} = 6.882$; yes, it does fit Po(2)

6 $\chi^2_7\ 5\% = 14.07$, $\chi^2_{\text{test}} = 13.18$; yes, they fit Po(2)

7 $\chi^2_5\ 10\% = 9.236$, $\chi^2_{\text{test}} = 9.80$; no, it does not fit N(16.2, 50.41)

8 $\chi^2_5\ 5\% = 11.07$, $\chi^2_{\text{test}} = 22.1875$; no, they do not fit N(62, 44)

9 $\chi^2_1\ 10\% = 2.706$, $\chi^2_{\text{test}} = 2.645$; accept H_0, die not biased

10 $\chi^2_1\ 5\% = 3.841$, $\chi^2_{\text{test}} = 3.39$; statement not supported

* 11 * (a)

	X	Y	
P	18	2	20
Q	14	16	30
	32	18	50

Expected frequencies

$$X \qquad Y$$

$$P \quad \frac{20 \times 32}{50} = 12.8 \quad \frac{20 \times 18}{50} = 7.2$$

$$Q \quad \frac{30 \times 32}{50} = 19.2 \quad \frac{30 \times 18}{50} = 10.8$$

O	E	$\dfrac{(O - E - \frac{1}{2})^2}{E}$
18	12.8	1.725 781 3
14	19.2	1.150 520 8
2	7.2	3.068 055 6
16	10.8	2.045 370 4

χ_1^2 1% = 6.635, χ_{test}^2 = 7.99; yes, they are dependent

(b) χ_1^2 10% = 2.706, χ_{test}^2 = 2.01; yes, independent

Chapter 9 Non-parametric tests

Basic

1
(a) $p(R \geq 5) = 0.2266$, which > 0.05, so accept H_0
(b) $p(R \geq 7) = 0.1719$, which > 0.05, so accept H_0
(c) $p(R \geq 8) = 0.0195$, which < 0.05, so reject H_0
(d) $p(R \geq 7) = 0.1719$, which > 0.1, so accept H_0
(e) $p(R \geq 8) = 0.0195$, which < 0.1, so reject H_0

2
(a) $p(R \geq 7) = 0.1719$, which > 0.05, so accept H_0
(b) $p(R \geq 9) = 0.0107$, which < 0.015, so reject H_0
(c) $p(R \geq 5) = 0.1094$, which < 0.11, so reject H_0
(d) $p(R \geq 6) = 0.0625$, which > 0.02, so accept H_0
(e) $p(R \geq 6) = 0.0625$, which > 0.04, so accept H_0

3
(a) $p(X \geq 6) = 0.1445$, which > 0.1, so accept H_0
(b) $p(X \geq 6) = 0.0625$, which < 0.1, so reject H_0
(c) $p(X \geq 7) = 0.1719$, which > 0.1, so accept H_0
(d) $p(X \geq 7) = 0.0352$, which < 0.05, so reject H_0
(e) $p(X \geq 9) = 0.073$, which > 0.05, so accept H_0

4
(a) (i) $X \geq 9$, $X \leq 1$
(ii) $X \geq 9$, $X \leq 1$
(b) (i) $X = 7$ or $X = 0$
(ii) $X = 7$ or $X = 0$
(c) (i) $X \geq 7$ or $X \leq 1$
(ii) $X = 8$ or $X = 0$
(d) (i) and (ii) $X \geq 12$ or $X \leq 3$
(e) (i) and (ii) $X \geq 8$, $X \leq 1$

5 $X \geq 10 = 0.1509$, which > 0.1, so accept H_0

6
(a) $Z = 2.014$, which < 2.326, so accept H_0
(b) $Z = 1.680$, which > 1.645, so reject H_0
(c) $Z = 1.610$, which < 1.96, so accept H_0
(d) $Z = 2.090$, which > 1.645, so reject H_0

7 $Z = 0.7606$, which < 1.960, so accept H_0; median is not different from 40

8 $Z = 1.896$, which < 2.054, so accept H_0; median is still 500 hours

9 $Z = 1.378$, which < 1.645, so accept H_0; median is not above national value

10
(a) (i) $Z = -0.594$, $|Z| < 1.960$, so accept H_0, they are from the same distribution
(ii) $U = 20$, so accept H_0
(b) (i) $Z = -1.408$, $|Z| < 1.96$, so accept H_0; they are from the same distribution
(ii) $U = 37$, so accept H_0
(c) (i) $Z = 1.733$, which < 1.96, so accept H_0; they are from the same distribution
(ii) $U = 36$, so accept H_0

*11 (a) G_1 18 19 24 25 27 29 30 32
     G_2 20 21 22 23 31
     R_1 1 2 7 8 9 10 11 13
     R_2 3 4 5 6 12

   $m = 5$, $n = 8$,
   $R = 3 + 4 + 5 + 6 + 12 = 30$
   $\frac{1}{2}m(m + n + 1) = \frac{1}{2}(5)(14) = 35$
   10%, one-tailed test, $Z_c = 1.282$

   $$Z = \frac{30 - 35 + \frac{1}{2}}{\sqrt{(\frac{1}{12} \times 40 \times 14)}} = -0.659$$

   Hence $|Z| < 1.282$, so accept H_0;
   there is no significant difference
   in the times.

   (b) $U = 40 + 15 - 30 = 25$, critical
       values for U are 8 and 32, so
       accept H_0.

12 (a) (i) $Z = -1.562$, $|Z| < 1.645$, so
            accept H_0, they are from
            populations with a common
            distribution.

        (ii) $U = 56 + \dfrac{7 \times 8}{2} - 42 = 42$.

            Critical values 10–46, so
            accept H_0.

    (b) (i) $Z = 1.173$, which < 1.645, so
            again accept H_0, their
            populations have the same
            distribution.

        (ii) $U = 24 + \dfrac{4 \times 5}{2} - 28 = 6$.

            Critical values 2–22 so accept
            H_0.

13 (a) $Z = 1.606$, which < 1.645, i.e.
       median is unchanged
    (b) $Z = 2.028$, which > 1.96, so yes,
       median is different

Intermediate

1 (a) $Z = 1.631$, which > 1.282, so
       reject H_0; second set better
    (b) $Z = 1.886$, which > 1.645, so
       reject H_0; second set better
    (c) $Z = 1.732$, which < 2.326, so
       accept H_0; no difference at 1%
    (d) $Z = 2.811$, which > 1.645, so
       reject H_0; second set better.

2 (a) $p(X \geq 7) = 0.172$, which > 0.05,
       so accept H_0; no significant
       evidence that wives are better
    (b) $p(X \geq 8) = 0.0547$, which > 0.05
       (just), so again accept H_0; no
       significant evidence that husbands
       are better, but they must have
       done some homework!

3 $p(X \geq 13) = 0.0106$, which < 0.02, so
   reject H_0; the median time is less.

4 $p(X \geq 9) = 0.0107$, which > 0.01, so
   accept H_0; the median has not
   increased.

5 $Z = 2.134$, which > 1.645, so reject
   H_0; median is lower than 65

6 (a) $Z = -1.035$, $|Z| < 1.645$, so
       accept H_0; second median not
       higher
    (b) $U = 22$, so accept H_0

7 $p(7 \text{ or more}) = 0.0352$, but only five
   show increases, so accept H_0

8 (a) $Z = 0.320$, which < 1.282, so
       accept H_0; there is no significant
       difference
    (b) $U = 24 + 10 - 24 = 10$, critical
       values 3–21, so accept H_0

9 $Z = 1.680$, which > 1.645 (just), so
   reject H_0, but M_2 is only slightly
   better

10 $Z = 1.483$, (5%) $Z < 1.645$, accept
   H_0; they have the same median
   (10%) $Z > 1.282$, reject H_0; forwards
   are faster

11 (a) $Z = -2.141$, $|Z| > 1.645$, so reject
       H_0; they have the same median
    (b) $U = 56 + 28 - 37 = 47$, critical
       values 13–43, so reject H_0; the
       three-quarters have a greater
       median according to this test

12 $Z = 0.350$, which < 1.645, so accept
   H_0; the median is not reduced

13 $Z = -0.423$, $|Z| < 1.282$, so accept H_0; the median is not less than claimed

*14 (a) $p(X \geq 8) = 0.0547$, which > 0.05, so accept H_0; there is no significant increase

   *(b)

2.5	2.8	3.4	3.1	3.5	2.7	2.4
2.4	2.6	3.6	3.1	3.6	2.9	2.6
−	−	+	0	+	+	+

2.6	2.9	3.2	3.3	2.8
2.8	3.2	3.3	3.5	2.9
+	+	+	+	+

$9 +$ signs, $2 -$ signs Bin$(11, \frac{1}{2})$
$p(X \geq 9) = (\frac{1}{2})^{11}(1 + 11 + 45)$
$= 0.0327$,

which < 0.04
So reject H_0, second set is significantly greater

15 (a) $Z = 1.071$, which < 1.645, so accept H_0; no significant evidence that price in T_2 is higher

   (b) $U = 42 + 21 - 41 = 22$, critical values are 8–34, so accept H_0

Advanced

1 (a) $Z = 1.606$, which < 1.645, i.e. accept H_0; there has been no improvement

---

2 $Z = 1.330$, which > 1.282, so reject H_0; numbers have declined

3 $Z = 2.897$, which > 2.326, so reject H_0; median mass is less than 500 g
$Z = 1.388$, which < 1.645, so accept H_0 at the 5% level but at the 10% level, when $Z_c = 1.282$, H_0 is rejected

4 $p(X \geq 8) = 0.0195$, which < 0.02, so reject H_0; there is improvement

5 $p(X \geq 7) = 0.1719$, which < 0.2, so reject H_0; they are heavier

6 $Z = 1.244$, which < 1.282 (just), so accept H_0; not justified to reduce time

7 (a) $Z = 1.400$, which < 1.645, so accept H_0; median time not more than 183 seconds

   (b) $Z = 1.750$. Critical values, 5%   1.645, 4%   1.751, so reject H_0 at the 5% level, and only just accept it at the 4% level

*8 (a)

Public house	1	2	3	4	5	6	7	8	9	10
Before	74	63	95	88	82	79	90	97	53	69
After	72	66	98	100	87	83	95	106	50	74
	−2	+3	+3	+12	+5	+4	+5	+9	−3	+5
	−1	+3	+3	−3	+5	+7	+7	+7	+9	+10

$P = 3 + 3 + 5 + 7 + 7 + 7 + 9 + 10 = 51$
$Q = 1 + 3 = 4 = T$

$P + Q = 55$
$\frac{1}{2}(10)(11) = 55$

1%, one-tailed test ... $Z_c = 2.326$

$Z = \dfrac{\frac{1}{4} \times 10 \times 11 - 4 - \frac{1}{2}}{\sqrt{(\frac{1}{24} \times 10 \times 11 \times 21)}} = 2.344$

$Z > Z_c$, but only just, so reject H_0. The median number of pints sold has increased. Further testing, however, is needed.

   (b) *Paired-sample sign test*: distribution is B$(10, \frac{1}{2})$

$p(X \geq 8) = (\frac{1}{2})^{10}(1 + 10 + 45)$
$= 0.05469$, or 5.5%

So reject H_0 at the 10% level, but accept it at the 5% level

9 (a) $Z = -1.299$, $|Z| < 1.645$, so accept H_0; there is significant evidence that the second crew does not have lower times than the first

  (b) $U = 35 + 15 - 24 = 26$, critical values 6–29, so accept H_0

Revision

1 $p(X \geq 9) = 0.0327$, which < 0.05, so reject H_0; median > 36 cm

2 $Z = 1.572$, which < 1.96, so accept H_0; median is not different from 454 g

3 $Z = 0.4226$, which < 1.645. so accept H_0; claim is accepted

4 $Z = 1.437$, which > 1.282, so reject H_0; mean is greater than 7

5 $p(X \geq 9) = 0.073$, which > 0.05, so accept H_0; median is not more than 51

6 $p(X \geq 8) = 0.0547$, so there is bias towards Tartarless for $p > 6\%$

7 $p(X \geq 10) = 0.898$, so no at 5%, yes at 10%

8 (a) $Z = 1.659$, which > 1.645 (just), so reject H_0; there is improvement, but more testing is needed

  (b) $p(X \geq 7) = (\frac{1}{2})^9 (1 + 9 + 36)$
$$= 0.0898,$$

  which > 0.05, so accept H_0; there has been improvement

9 A 31 33 30 28 34 32 28 29 35 27
  B 32 34 28 31 38 35 24 31 40 33
  +1 +1 −2 +3 +4 +3 −4 +2 +5 +6

Ranks
+1.5 +1.5 +3.5 −3.5 +5.5 +5.5 +7.5
−7.5 +9 +10

$P = 1.5 + 1.5 + 3.5 + 5.5 + 5.5 + 7.5$
$\quad + 9 + 10 = 44 \qquad\qquad n = 10$

$Q = 3.5 + 7.5 = 11 = T$
$P + Q = 55;\ \frac{1}{2}n(n+1) = \frac{1}{2}(10)(11)$
$$= 55$$

For a one-tailed, 5% test, $Z_c = 1.645$

$$Z = \frac{\frac{1}{4} \times 10 \times 11 - 11 - \frac{1}{2}}{\sqrt{(\frac{1}{24}) \times 10 \times 11 \times 21}} = 1.631,$$

$Z < Z_c$ (just)

So accept H_0; there has been no significant increase, but more testing is needed.

10 $U = 30 + \dfrac{5 \times 6}{2} - 26 = 19$, so accept H_0; there is no significant difference

Printed in Great Britain
by Amazon